American Fire Apparatus Vol. 1 - Pumpers

by Wayne Mutza

Covers and art by Don Greer
Line Illustrations by Matheu Spraggins

Squadron/Signal Publications

(Front Cover) Despite the popularity of the classic American LaFrance pumper, the Milwaukee Fire Department has operated only one since 1961. This 1973 Century Series was the department's first pumper capable of pumping water at a rate of 1,500 gallons-per-minute. Working at a warehouse fire, Engine Company 1 draws water from a hydrant while discharging water to supply hand lines and its deck pipe.

(Back Cover) King among pumpers was the Fire Department of New York's Superpumper System, placed in service during 1965. The system used five massive Mack-built vehicles, including Super Tender 1, which was the companion to Super Pumper 1. When supplied with water by the Superpumper, Tender 1's 10,000-GPM monitor could throw a powerful stream of water at a distance of 600 feet.

Copyright 2009 Squadron/Signal Publications
1115 Crowley Drive, Carrollton, TX 75006-1312 U.S.A.
Printed in the U.S.A.

All rights reserved. No part of this publication may be reproduced, stored in a retrieval system, or transmitted in any form by means electrical, mechanical, or otherwise, without written permission of the publisher.

ISBN 978-0-89747-593-8

Military/Combat Photographs

If you have any photos of aircraft, armor, soldiers, or ships of any nation, particularly wartime snapshots, why not share them with us and help make Squadron/Signal's books all the more interesting and complete in the future? Any photograph sent to us will be copied and returned. Electronic images are preferred. The donor will be fully credited for any photos used. Please send them to:

Squadron/Signal Publications
1115 Crowley Drive, Carrollton, TX 75006-1312 U.S.A.
www.SquadronSignalPublications.com

About the Special Series

Squadron/Signal Publications' most open-ended genre of books, our Special category features a myriad of subjects that include unit histories, military campaigns, aircraft, ships, armor, and uniforms. Upcoming subjects include war heroes and non-military areas of interest. If you have an idea for a book or are interested in authoring one, please let us know.

Acknowledgments

This book would not have been possible without assistance from Christopher Allen; Scott W. Anderson, Oshkosh Truck Corp.; Dick Bartlett; Robert Brackenhoff; Peter Byrne; John A. Calderone, *Fire Apparatus Journal*; Patrick Campbell; Bill Friedrich; Jeff Holter; Richard "Rick" Howard; Sarah Judd, New York City Fire Museum; Janine Kozak; Scott A. LaPrade; Mike Legeros; Dennis J. Maag; Chuck Madderom; Gerrit Madderom; John Peter Maguire; Matt J. Russ II; Tom Shand; Shaun P. Ryan; Richard Schneider; Patrick Shoop, Jr.; Mark Stampfl; Jim Stevenson; Robin Walker, S & S Fire Apparatus; Cliff Williams; and Eric Zak.

Dedication

To the Fire Department of New York's 343 firefighters whose last response was to the World Trade Center on 11 September 2001.

(Title Page) All Engine companies of the Fishers, Indiana, Fire Department are called "Advanced Life Support" (ALS) engines since each vehicle carries a paramedic among a crew of four. Fishers' Engine 93 is a 1998 KME "Excel" 1,500-GPM pumper with a 750-gallon water tank and 50 gallons of foam. Engine 93 wears a fire prevention message on its high-visibility body. (Christopher Allen)

Introduction

Throughout history, as society endeavored to exploit the usefulness of fire, devising ways to counter fire's destructiveness also became a necessity. Since water was the most obvious and available means of quelling the ravages of fire, how the water was applied depended upon the ingenuity of pioneers of fire extinguishing equipment.

After the first English settlement was established at Jamestown, Virginia, other American settlements grew and teemed with wooden structures, often built close to each other. With open flames used in abundance for heat and light, dwellers grew fearful of the cry "Fire!" In Colonial America, the first organized efforts to combat fire required residents to possess leather buckets. When the alarm of "Fire" rang out, it was followed by "Throw out your buckets!" After scrambling to collect their buckets, volunteers formed "bucket brigades," which strung from a water source to the fire, and passed the buckets to the front of the line where the bravest fought the flames. Against the enemy fire, courage was not enough, however, and from the beginning, mechanical tools were developed to throw more water farther, to climb higher, and to respond to any emergency.

Soon, organized fire protection was addressed in the growing harbor communities of New York, Boston, and Philadelphia. As communities grew, fire fighting techniques and equipment were developed to keep pace with the ever-increasing hazards of fire. A common misconception is that original fire fighting equipment was pulled by horses, when, in fact, it was pulled by hand. The first wheeled equipment consisted of large water containers fashioned by craftsmen, who mounted the containers to wagon wheel carriages. Such "tubs" became hand pumpers, with the addition of two long parallel handles called "brakes." Scores of men pumped the brakes to force water under pressure through a hose. The effectiveness of these rudimentary machines was limited by the volunteers' endurance. The first hand pumpers used in America were built in England and shipped to New York during the early 1700s. Soon, craftsmen in America were turning out hand pumpers for mushrooming communities across the country.

The first volunteer firemen were a bawdy, colorful lot who played as hard as they worked. True to their male machismo, they viewed fire fighting as personal combat and embraced their status in society as a noble breed. As volunteer firemen proved themselves to be the ablest and the bravest, fire fighting in Colonial America took on a romantic flair. Their elite ranks included men of prominence such as John Hancock, Paul Revere, George Washington, and Benjamin Franklin. Although Boston and New York, along with other cities, had organized fire watches during the mid-1600s, Franklin was credited with forming, in 1736, one of the first official fire fighting organizations—Philadelphia's Union Fire Company. The volunteers' status was embellished with ballads written about their exploits. Their wheeled equipment, thereafter called "apparatus," was specially painted and bore individual names. Placing "first water" on the fire became a matter of utmost importance and pride. Competition among the volunteer companies was fierce, and rivalry abounded. At times, the fire was ignored as companies brawled, bombarded each other with hose streams, or sabotaged each other's equipment. Often, pails of beer or stronger libations were brought to blazes as volunteers waged war against the flames.

As fire companies were organized, their place in society was so prominent that they became the only civil authority and even quelled labor and ethnic riots in major cities. From these duties, the nation's first police organizations arose. Out of such traditions of pride, bravado, and teamwork, fire departments across America were born.

As steam power was developed during the early 1800s, steam pumpers proved capable of providing a continuous stream of water under pressure with far less manpower. Although the first steam pumpers were pulled by hand, volunteer firemen saw them as a threat to their existence. Firemen proved, during demonstrations, that they could pump more water by hand than early smoke-belching steam contraptions, but soon mechanical power won over muscle power. As the need for horses to pull heavy steamers became evident, volunteers in major cities were replaced by paid fire departments. During the rapid growth of major cities in America, the destruction wrought by huge blazes taught fire departments the painful lessons of having too few fire-fighting vehicles and the need to improve them. Such devastation also spawned stricter fire codes and ordinances.

When the motor car revolutionized transportation in America at the beginning of the 20th Century, horses were slowly and often grudgingly phased out. Since municipalities could not afford complete conversions to motorized apparatus, their two primary and largest types of apparatus, steam fire engines and ladder trucks, were pulled by motorized tractors. Despite experimentation with electric and steam-powered vehicles, gasoline would prevail in replacing horses. Smaller pieces of single- and dual-purpose equipment, such as hose wagons and chemical units, which previously had been pulled by horses, were built onto single automotive chassis. Incorporating multiple functions into one vehicle resulted in the "triple combination" pumper. Such cost-saving and manpower-reducing measures eased the transition from horses to horsepower as motor transportation swept the nation.

As fire departments struggled to keep pace with growth, technology, and tight budgets, some auto and truck manufacturers capitalized on their needs by making their proven commercial chassis available to firms that specialized in fire apparatus. Although such companies sprang up from coast to coast, the majority of fire apparatus was built by companies that built the entire vehicle on their own custom chassis. Eventually, both urban and rural fire departments were ordering emergency equipment tailored to their special needs. Such needs and the imagination of engineers and designers resulted in a wide and interesting variety of fire apparatus, many of which were designed to handle emergencies other than fire.

Faced with ever-present budget restraints, some fire departments proved their resourcefulness by building their own apparatus and associated equipment. Talented craftsmen in some cities built apparatus from the ground up while others updated apparatus, often completing multiple conversions of the same rig over many years. Fire departments, more than any other organization, have become known for their ability to extend the life of their vehicles.

The innovations and improvements made to fire apparatus over the years set them apart from vehicles that traveled the city streets and country roads of America. Firms that built their own custom fire apparatus used signature features that identified their product. Besides their unique purposeful designs, fire apparatus was marked by the traditional color—red. These volumes, however, show that they were not limited to that color. Red was first used by the volunteers of Colonial America not only to mimic the color of fire itself, but to proudly display their apparatus at competitions. In addition, history records that when Henry Ford began producing motor cars, they were available only in black, prompting the use of red to give fire apparatus visibility among throngs of black vehicles.

The alarming rate of firefighter deaths and injuries as a result of traffic accidents, especially

This hose reel was built during the 1850s for the Philadelphia Fire Department. Iron scrollwork was typical of the craftsmanship of early fire apparatus. Lying between its wheels is the detachable pole with iron hand pulls. (Mike Legeros)

The Arrow Company of England built this hand pumper, the rotating pump handles of which are shown in the stowed position. Side panels of water tanks and condenser boxes often were decorated with patriotic symbols or portraits of government leaders. (Mike Legeros)

at intersections, led to studies during the 1960s of alternative colors for fire apparatus. These studies found that yellow and green-yellow were colors to which the eye is most sensitive. After additional research by medical professionals and safety agencies, fire departments across the country began switching to apparatus painted the green-yellow livery, called both "lime-yellow" and "lime-green." For a time, more fire trucks were leaving the production line painted this high visibility color rather than red. Even the world's largest fire department, which protected New York City, experimented with lime-yellow, beginning in 1980.

Despite the broadened safety margin afforded by lime-yellow fire apparatus, the color was unpopular with both firefighters and the general public. Tradition runs deep in the fire service, and most departments reverted to red apparatus although many followed the example of New York, which switched to red and white in 1984. Most departments that changed back to red relied on reflective trim and eye-catching light systems to enhance apparatus visibility, especially at night. Lime-yellow, however, remained the standard for airport and military fire apparatus.

This series of three volumes, a significant deviation from Squadron/Signal's line of books covering military hardware, takes a look at the fascinating world of American fire apparatus; from the rudimentary equipment of Colonial times to today's high-tech vehicles; from the fleets of major city fire departments to the smallest volunteer companies. This first volume covers what has continually been considered the central front-line fire fighting vehicle—the pumper.

The time-honored purpose of the pumper, which is more commonly called an "engine" in fire service jargon, is to pump water from a source and discharge it under pressure through hoses to fight fire. Over time, that basic concept has evolved to the extent that engines often resemble buses that are massive wheeled pumping stations, which not only carry their own large supply of water, but also perform multiple operations. The engines do more with less in the interest of saving costs and manpower.

The development of some multi-purpose fire apparatus was so extensive, their pumps became simply one of many features. It is not uncommon for engines to carry the label of three or four manufacturers, whose major components were used to construct the vehicle. Pumpers, typically, are identified by their manufacturer and pump rating, which is measured in gallons per minute (GPM), usually in increments of 250 GPM, the standard rate of one discharge outlet. Pump ratings range from 100 GPM to 2,000 GPM, or even higher in the case of massive pumper apparatus, such as the Superpumper System. Descriptions of pumpers may also include total capacities of water and foam tanks as well as other extinguishing agents.

Following this present first volume on the pumper, Volume Two will be devoted to an examination of all forms of aerial fire apparatus while Volume Three will provide insight into the incredibly wide range of specialized apparatus outside the realm of pumpers and aerials. Although fireboats are often categorized as marine engine companies, they will be covered in Volume Three.

Despite the passage of time, the fire service remains steeped in tradition. While much of the romance of fire fighting from the days when "fire laddies" raced down cobblestone streets may be gone, many of the basic concepts and even the terminology of those days lives on. No matter how advanced we become, fire will always be with us as an "equal opportunity destroyer," requiring fire apparatus that captivates our interest and admiration. Who doesn't stop and turn to watch a massive, gleaming fire truck roaring down the street and bearing firefighters who make their living facing the flames?

The Early Years

Bethlehem, Pennsylvania, claims to have taken delivery of the oldest piece of fire apparatus in the U.S. Built in London in 1698 by Brooks, the hand pumper arrived on the ship *Hope* on 21 October 1763, and was delivered to Bethlehem on 10 December 1763. Bethlehem ordered the apparatus to protect more than 30 industries that prospered in the thriving settlement.

As cities grew during the early 19th Century, volunteer firemen found that they could not rely on a few hand-pulled and operated "tub" pumpers to control fires that swept through rows of buildings. Fortunately, steam pumpers were being developed in England. Although Swedish inventor John Ericsson built the first American steam-powered fire engine in 1840, they didn't begin to appear in America until Alexander Latta, Abel Shawk, and Miles Greenwood of Cincinnati, Ohio, built the first practical steam fire pumper. Named the "Uncle Joe Ross," it was presented to the Cincinnati Fire Department on 1 January 1853. Their steamer could throw a stream of water 240 feet, which rendered hand pumpers obsolete.

Early steam pumpers were hand-drawn but quickly grew in size, and they had to be pulled by horses. Since steamers normally were built only with riding provisions for an engineer, foreman, and pipeman, other firemen had to run alongside it. Soon, horse-drawn hose carts, or hose wagons, carried hose and firefighters and accompanied steam pumpers. The switch to steamers, horsepower, and the need to train men to operate the machinery spelled the end of volunteers in major cities and the beginning of organized, paid fire departments. Boston took the lead in making fire-fighting a profession in 1678. New York, on the other hand, held out until 1865 before establishing a paid fire department.

Among the first steamer manufacturers were Nott, Corbett, Metropolitan, Ahrens, American, La France, Silsby, Gould, Button, Amoskeag, Clapp & Jones, Lee & Harned, and Harrell. Silsby Manufacturing Company of New York was the oldest firm, having built more than 1,000 steam fire engines during the steam era. In 1891, Silsby merged with Ahrens, Clapp & Jones, and Button to form the American Fire Engine Company. Although their designs differed, their operation was basically similar. Water in the steamer's pressure vessel, called a boiler, was heated with coal to produce steam. Steam pressure then powered double-acting, two-cylinder pump engines. Pumps were regulated so that water could be drawn from a source and then through leather, nozzle-tipped hoses manned by firefighters. Typical operation of a steam pumper called for water in the boiler to be kept hot by a small steam coil connected to a steam generator in the firehouse. The engineer kept kindling, often soaked with lard or kerosene, on the grates below the boiler, at the rear of the unit. When the alarm sounded and as soon as the steamer's stack cleared the doorway, the kindling was torched to ignite a pile of coal. Normally, within five to ten minutes, a head of steam sufficient to provide pump pressure would be produced. Enough coal was carried to produce steam for about 30 minutes. At large fires, a coal wagon supplied coal, which enabled "stokers" to continuously fire boilers. Later, back at the firehouse, the engineer dumped his fire on the ground, cleaned the combustion chamber, and laid fresh kindling on the grates.

Steamers were categorized by their pumping capacity in gallons per minute (GPM), a practice that continues with today's pumpers. Seven steamer types existed, all of which measured approximately 9 feet in height and between 22 and 25 feet in length. Depending on size, steamers weighed between 4,000 and 8,500 pounds. The following categories are steamer types:

- Extra First Size – 1,100 GPM
- First Size – 900 GPM
- Second Size – 750 GPM
- Third Size – 650 GPM
- Fourth Size – 550 GPM
- Fifth Size – 450 GPM
- Sixth Size – 375 GPM

The glamour and appeal of the intricate steam pumper, with its gleaming stack, often overshadowed the heart of the system—the firehorse. Horses were the pride of the fire department, and they were revered by their fire-fighting handlers, who considered the animals hard-working, faithful, and intelligent. Horses also had their own personalities, and it was not unusual for them to retaliate against those who treated them poorly. Veterans of the horse-drawn era recounted how horses perked up their ears and kicked their stalls when the alarm sounded for their "first-in" area. When released from their stalls, horses dashed to their spots below harnesses suspended from elaborate trapeze devices. Split collars dropped onto their necks while bits and reins were quickly snapped together. In less time than it took for crews to climb aboard, horses were hitched, and the harness contraption swung out of the way.

Steam pumpers usually were pulled by a three-horse hitch, which was said to have seen its first use by the Fire Department of New York in 1884. Less common was a "spike hitch" in which a more seasoned horse was hitched ahead of two abreast. Smaller vehicles, such as hose wagons and chemical carts, were pulled by two horses, while chief's buggies required only one horse. On the run, engineers holding the reins maintained careful balance and control of the powerful animals in case they stumbled or the apparatus overtook them. At the fire, horse teams were unhitched and were either walked or kept in holding areas. Fire departments often had their own veterinary physicians, who attended major fires to look after the horses. Large cities not only built shops for mechanical apparatus, but they also built horse hospitals.

As buildings rose above two stories, the second major type of fire apparatus, the ladder truck, was also pulled by teams of horses. Fire departments quickly learned that other types of specialized apparatus were needed to support steam pumpers and ladder trucks. For smaller fires and to quickly attack blazes while steamers prepared to go into operation, the chemical unit was developed. Pulled by horses, chemical units carried tanks of bicarbonate of soda, which, when activated by sulfuric acid, created pressure to discharge the chemical through small-diameter rubber hoses. Combination hose wagon chemical units became popular, and eventually, the chemical extinguishing option was added to ladder trucks.

Cotton-jacketed hose was developed during the 1880s to replace bulky leather hose and rubber hose carried in baskets or on reels. Large hose wagons not only could carry more of the flat-laid hose, but they also often mounted large stream appliances, which were called "deck pipes." These "turret wagons" could be positioned close to large fires to enable the deck pipes to attack blazes with high pressure streams.

With the advent of motor power after the turn of the century, horsepower inevitably would replace horses. The changeover, however, would not come easy, since fire departments were resistant to change and found their horses more reliable than complex vehicles that were subject to mechanical breakdown. Easing the transition was the fact that horse-drawn fire apparatus had grown so large and heavy that horse teams became exhausted after short, high-speed runs. Such short distances required more firehouses in expanding communities. During the changeover, which lasted two decades, the horse-drawn apparatus was improved by replacing steel wheels with rubber wheels while hub brakes replaced rim, or block, brakes.

The Radnor Fire Company of Wayne, Pennsylvania, claims to have acquired the first motorized fire apparatus in the U.S. Radnor placed the first order with the Knox Automobile Company at Springfield, Massachusetts, in 1906. The Thompsonville, Connecticut Fire Department, however, had also placed an order and received the "first" Knox. The vehicles were equipped with two gasoline motors. One propelled the vehicle to its top speed of 25 mph while the other powered the pump. The pumpers carried 35 gallons of water and 1,000 feet of hose. Three Radnor firefighters became so experienced with the pumper, they formed the Hale Fire Pump Company. After shifting operations to Conshohocken, Pennsylvania, Hale became a mainstay in the fire apparatus industry. The Fire Department of New York (FDNY) received its first gasoline-powered pumper in 1909, and 20 December 1922 was the last time that a team of horses made its run pulling an engine through New York streets.

Although single-chassis motorized fire apparatus began appearing in fire departments, the horse teams of steam pumpers and ladder trucks were commonly replaced by two-wheel tractors. By 1930, horses no longer appeared on fire department inventories. With motorization came the triple-combination pumper that incorporated a pump, a hose, and chemical equipment on a single chassis. The first example was built in 1909 by the Tea Tray Company of New Jersey and was delivered to Middletown, New York. The popular chemical extinguishing unit on triple combinations would be replaced by a "booster reel" of small-diameter rubber hose, which used water pumped from an onboard tank.

(Above) Some of the first successful hand pumpers were built by Richard Newsham of Cloths Fair, England, beginning in 1721. New York City took two, in 1731. The Newsham held 150 gallons of water and pumped 100 gallons of water per minute to throw a stream from its nozzle atop the unit at 75 feet. Lettering on this engine, which is constructed of wood and iron, states "Germantown 1764." It is displayed at Philadelphia's Fireman's Hall Museum. (Mike Legeros)

(Right) This end-stroke, Philadelphia-style hand pumper is one of about 20 built by the partner firm Mason & Gibbs. Like all hand pumpers, its tank was filled by a bucket brigade as water was ejected through the hose/nozzle fitting atop the condenser box. Markings include the date 1792. (Mike Legeros)

This unusual hand-drawn pumper required two operators to work the long handles in a see-saw fashion that activated its twin-cylinder pump. (Mike Legeros)

Until the arrival of larger, horse-drawn fire apparatus, two-wheel hose carts were hand-pulled to fires by companies, which typically comprised nine men. This "Jumper" was one of Milwaukee, Wisconsin's, early pieces of fire apparatus. (Author's Collection)

Built with an abundance of brass, this early Pioneer steam pumper was used by the Insurance Patrol. A buckboard-style seat was mounted above the single hose discharge fitting. (Mike Legeros)

The glamour and excitement of a steamer racing to a fire was matched only by the phenomenon of the fire itself. On the run is Engine Company 9, the oldest company in the New York City Fire Department, which traces its history back to the Colonial days of the late 1700s. When this photo of Engine 9 was taken on 10 December 1911, the company ran out of its Manhattan firehouse on East Broadway, where it was stationed for more than 100 years. Following close behind is the steamer's wagon, which carries hose and manpower. (Author's Collection)

Pittsburgh's Bureau of Fire, Department of Public Safety patented this hanger frame for hitching a two-horse team to Turret Wagon 2. Attached to the apparatus, this rig gave the flexibility to park the wagon anywhere in the firehouse. (John Norman McIntyre Collection)

In 1872, in the midst of the American Industrial Revolution, Truckson LaFrance and his partners started the LaFrance Manufacturing Company at Elmira, New York, to produce hand pump and steam-powered fire apparatus. At the turn of the century, the firm joined with the American Fire Engine Company to become the famed American LaFrance Company. (American LaFrance)

During the early years, when an alarm sounded in the firehouse, conditioned horses were released from their stalls and positioned themselves under elaborate trapeze rigging where they were quickly hitched to collars and harnesses. This hose wagon of Protective Company 1 of Rochester, New York, used a split-collar system for a two-horse hitch. (John Norman McIntyre/Gus Johnson Collection)

Engine Company 2 of Pittsburgh's Bureau of Fire, Department of Public Safety was a chemical engine that mounted a large turret. When fed by a large-diameter hose, the turret could direct a powerful stream of water. Small front wheels increased the rig's turning radius. The company mascot sits atop the chemical hose reel. (John Norman McIntyre Collection)

The driver of Engine Company 13 of the Buffalo, New York, Fire Department pulls hard on the reins to control his three-abreast hitch. (Author's Collection)

Buffalo, New York's, Hose Company 23 runs closely behind a steam pumper as its officer, seated next to the driver, hurriedly dons protective clothing. (Author's Collection)

The Perkins Company of Lawrence, Massachusetts, built this chemical unit for the Lynn, Massachusetts Fire department. Chemical tanks were usually constructed of heavy rolled copper and lined with lead to minimize the destructive action of carbonic acid produced by the mixing of bicarbonate of soda with sulfuric acid. The latter was kept in a glass container built into the mixing tank. (Author's Collection)

One of the first Christie front-drive tractors was used to pull this steamer, belonging to Community Fire Company 1 of Wayne, New Jersey. Compared to Christie competitor American LaFrance, the steering column was vertically mounted. Boiler stacks, bells, and other gleaming components were nickel-plated. (John Peter Maguire)

The central fire station of Houston, Texas, was considered the most elaborate in the country. Built in 1903, the architectural masterpiece served for 20 years before it was razed. Shown in 1915 is the array of horse-drawn and motorized apparatus housed in the eight-bay structure. (Houston Fire Department)

An early experiment with mating steam pumpers to a motorized unit was this rig of the Fire Department of New York (FDNY). The driver sat atop the motor, which provided front drive. (Author's Collection)

The Cambridge, Massachusetts, Fire Department placed this First Size steam pumper in service in 1872. In 1919, Engine Company 3 was converted from horses to horsepower with a four-cylinder, front-drive American LaFrance tractor. (Author's Collection)

Fire departments in regions that endured harsh winters often relied on skis to get apparatus to a scene. The Milwaukee Fire Department used this long sled to haul frozen hose from a fire scene. A ratchet gong is mounted next to the driver's seat. (Author's Collection)

The Christie front-drive tractor was probably the most popular type used to replace horses. This 1914 Christie was attached to Engine 1's steamer of the Portland, Oregon, Fire Department. A tool box is mounted above the front wheel. A pull cord is visible, and it runs from the bell to a rear riding step, where a firefighter rang the bell. (Rick Howard Collection)

The Hutchinson, Kansas, Fire Department used this two-cylinder-powered, 1906 Rapid chemical and hose car. The Rapid Motor Vehicle Company of Pontiac, Michigan, was founded in 1902 and was purchased by General Motors in 1909. It 1912, it became GMC Truck. Chains on the solid rubber tires increased traction on mud and snow-covered streets. (Author's Collection)

Motorization

Throughout the development of motorized fire apparatus, the pumper's basic operation remained unchanged. Pumpers draw water from a source, often pressurized hydrants, and through a suction hose mounted low on the pumper's frame. "Hard suction hose," which is semi-rigid to prevent collapse while drafting from open water sources, is carried should hydrants be unavailable. Water tanks, called "booster tanks," built within the pumper's body allow firefighters to quickly attack a fire, or to extinguish small blazes. As additional pumpers arrive, they tap into unlimited water sources, relaying water to the first-arriving engine company. The driver/pump operator's work station is the pump panel, which most often is located on the pumper's left side. Pump operators rely on the panel's gauges to monitor the pump's operation as water courses through discharge gates into hoses. Various types of large stream appliance mounted to the apparatus, such as deck pipes, turrets, or nozzles incorporated into aerial ladders or articulating booms, are fix-piped directly to the pump.

The most popular front-drive tractors used by fire departments to pull steam pumpers were built by American LaFrance and Front Drive Motor Company (Christie). Seagrave also established itself in the field of motorization, building both two-wheel and four-wheel tractors, with the latter commonly used to pull ladder trucks.

In 1911, the Ahrens Fox Fire Engine Company introduced its unconventional pumper with the pump mounted forward of the engine, and two years later introduced a booster car that had a small pump, a water tank, and a hose. The design soon replaced chemical engines, and the booster system became standard on triple-combination pumpers. Many departments continued to use chemical sub-systems on apparatus into the 1930s.

A hand-cranked siren appeared in 1913, although bells remained on apparatus more as tradition than function. During the 1920s, shaft-driven power trains replaced chain-drive. Apparatus was equipped with right-side or left-side steering, but the latter became standard by the end of the decade. Cabs were left open and were often without doors or windshields to allow visibility and egress when arriving at a fire scene. Although well ahead of its time, Pirsch, in 1928, built what is considered the first enclosed custom-built cab.

During the 1930s, engines began carrying their own water; windshields became standard; and pneumatic tires replaced solid rubber wheels. The decade also saw the introduction of one of the most revolutionary designs in fire apparatus—the cab-forward chassis. Positioning the engine in the rear portion of the cab not only increased visibility, it also improved turning radius and provided protective riding positions for firefighters. In 1937, American LaFrance, a leader in the fire apparatus industry, revealed its cab-forward design, which would be adopted by most custom apparatus builders.

The quad arrangement, which incorporated pump, booster or chemical tanks, hose body, and ground ladders, was not only improved during the late 1930s, but it also became a quint with the addition of an aerial ladder. These multi-function vehicles appeared in some large cities to minimize manpower and costs, but more often were found in smaller communities where large numbers of conventional apparatus were unnecessary. The number of fire apparatus builders rose steadily through the 1930s and filled orders for both conventional and specialized apparatus.

John Walter Christie was known more for his work with military hardware and race cars than front-drive fire apparatus tractors, which he began building in 1912. Yet, nearly 600 Christies were sold through the Front Drive Motor Company of Hoboken, New Jersey. The Fire Department of New York began motorizing its fleet of steam pumpers with Christies, of which the first 16 arrived in 1916. Engine 93, seen here, which originally ran with two hose wagons in lower Manhattan, is now displayed in the New York City Fire Museum. (Author's Collection)

Engine 3 of the Portland, Oregon, Fire Department used this 1917 American LaFrance front-drive tractor to pull a steamer using a "goose-neck" frame. The tank beneath the seat held 20 gallons of gasoline, which was gravity-fed to a 75 horsepower, four-cylinder motor. Front wheels were cast steel discs with solid rubber tires. The 12-inch, electric headlights were a far cry from the hand lanterns previously used. (Author's Collection)

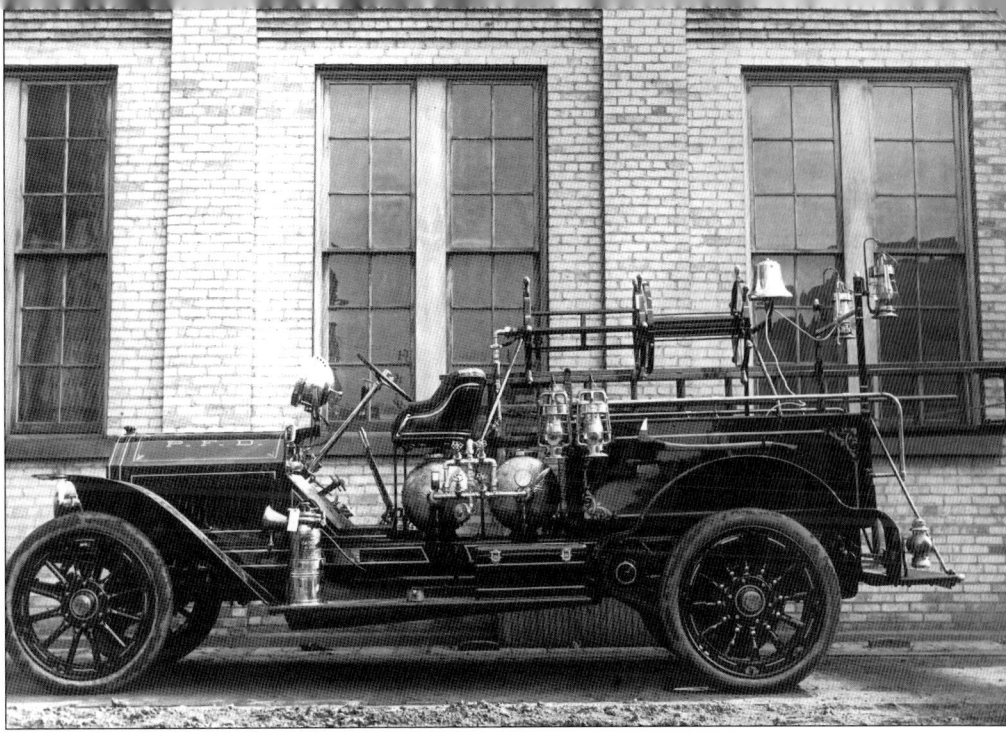

This 1917 Howe is credited with being the first motorized "automobile" type of apparatus in the state of Maryland. Continuously operated since 1818, Frederick's Independent Hose Company 1 is the oldest fire company in Maryland. This pumper had a rear-mounted pump. The ungainly ladder rack was later removed. (John Norman McIntyre Collection)

This 1913 American LaFrance Type 10 combination engine went into service as Engine 23 of the Portland, Oregon, Fire Department. The rear-chain drive unit was equipped with a hose bed, ladders, and twin chemical tanks with hammered ends. At least six hand lanterns were carried on the rig. The twin hose reels ringed with rewind handles had yet to be outfitted with rubber chemical hose. (Rick Howard Collection)

The E.H. Stokes Company of the Volunteer Fire Department of Ocean Grove, New Jersey, ran this chemical car as Engine Number 3. Using equipment from a horse-drawn chemical unit, the Autocar Company of Buffalo, New York, built the unit in 1909. Attached underneath the custom rear running board is a ratchet gong, which was rung by a foot pressure switch. (Author's Collection)

This 1916 American LaFrance Type 12 with T-head engine served as Engine 26 of the Milwaukee Fire Department. The 750-GPM pumper was one of three purchased by Milwaukee. Two suction hoses were normal on Milwaukee rigs to take advantage of the city's ample cisterns and inland waterways. (Author's Collection)

Shelbyville, Tennessee, operated this 750-GPM 1926 Mack AC-2 pumper until the 1950s. The white-painted rig featured enclosed chain drive. A metal extension ladder was added by the department. Called the "Bulldog" because of its toughness and blunt snout, the AC model was introduced in 1916 and became available in 13 types of fire apparatus. (Author's Collection)

After starting the South Bend Motor Car Works in 1912 to build fire apparatus, Alfred C. Mecklenburg joined Harry C. Stutz in Indianapolis to form the Stutz Fire Engine Company. Called "Old Engine 1," this 1937 Stutz was the first diesel-powered fire apparatus built in the U.S. The pumper was taken on a promotional tour before going into service as Engine 1 with the Columbus, Indiana, Fire Department in September 1939. Its original engine was a Cummins six-cylinder in-line supercharged model HRS-6 diesel. Engine 1's 1,000-GPM rotary-gear pump, 100-gallon water tank, and 1,500 feet of 2 1/2-inch hose made it a triple combination pumper. (Christopher Allen-Indiana Fire Trucks.com)

The Cambridge, Massachusetts, Fire Department ran this American LaFrance Type 40 combination chemical pumper as Engine 3 from 1920 to 1950. The truck was four-cylinder-powered and had a 250-GPM pumping capacity. It was equipped with tanks of bicarbonate soda and water, to which sulfuric acid was added at the fire scene to create carbon dioxide gas to force water from the tank. (John F. Collins)

In 1948, Engine 1 was re-powered with a Cummins naturally-aspirated HR-6 engine. It was retired in November 1974 and was restored, in 1989, by Cummins Engine Company to commemorate the firm's 50th anniversary. (Eric Zak)

This 1916 Seagrave chemical and hose car belongs to the Reno, Nevada, Fire Department. Leather-upholstered seats, tool boxes on side running boards, and oval or elliptical fuel tanks mounted behind the seat were common features of apparatus from this era. (Shaun P. Ryan)

The FDNY used this 1929 Seagrave hose wagon to accompany Engine 81, which had been organized in the Bronx in 1913. Grab rails surround its hose bed and seat. The inlet for supplying the turret pipe protrudes below the cabinet behind the driver's position. (Author's Collection)

This 1930s Dodge Brothers pumper has extra long suction hoses, called "squirrel tails," that ensured reach to a water source and were usually stored wrapped around the front of the truck. (Shaun P. Ryan)

This 1921 American LaFrance Type 75's low weight and maneuverability made it a popular fire truck. Power was supplied by a six-cylinder T-head engine, with a 750-GPM rotary-gear pump. (www.fireapparatusphotography.com)

This 1923 Stoughton-built pumper was rated at 350 GPM and carried a 200-gallon water tank. It served the Antioch, Illinois, Fire Department, and later the Lauderdale-Lagrange Fire Department of Whitewater, Wisconsin. (www.fireapparatusphotography.com)

This 1923 Ahrens Fox as Engine 12 has a water tank mounted immediately behind the fuel tank atop the rig. The classic Ahrens Fox is easily identified by its front-mounted pump with a large round chamber. (Bruce Neal/The Antique Fire Brigade Collection)

Cokato, Minnesota's, first motorized apparatus was this 1928 Reo. (Shaun P. Ryan)

This vehicle is a 1934 Chevrolet chemical truck .(www.fireapparatusphotography.com)

This 1929 American LaFrance pumper served the Kearney, Nebraska, Volunteer Fire Department. Spotlights were common fixtures on the forward part of cabs. (Shaun P. Ryan)

This 1924 American LaFrance (ALF) was ALF's answer to a low-priced rig for small departments. The ALF was built on a Brockway "Torpedo" chassis. The pumper's four-cylinder, Wisconsin Motors engine also powered a 300-GPM rotary pump. The rig featured an 80-gallon water tank. (Chuck Madderom)

Complete with a brass eagle atop its dome and a Roto-Ray light behind the cab, this 1937 Ahrens Fox belonged to the Town Bank Fire Department of Cape May County, New Jersey. (Shaun P. Ryan)

This vehicle is a 1930 Ahrens Fox named "Rosie" of the Keene, New Hampshire, Fire Department. (Dick Bartlett)

Maxim apparatus, such as this 1935 model of Marion, Massachusetts, were prominent along the East Coast. (Dick Bartlett)

Reo, long a big name in the truck building industry, built this 1930 model wagon for Petaluma, California. (Shaun P. Ryan)

Resplendent in its gold leaf trim and brass fittings, this 1937, 500-GPM Seagrave combination pumper was owned by the fire department of San Bruno, San Mateo, California. (Shaun P. Ryan)

This vehicle is a chain-driven, 1926 Mack AC of the Calais, Maine, Fire Department. A rendition of Mack's popular bulldog was painted on the white water tank. Dual rear wheels were uncommon during the 1920s. (Dick Bartlett)

Equipped with a Hirst 200-GPM pump, this 1937 Studebaker ran with the Georgetown Volunteer Fire Department of El Dorado, California. Although built on an average size chassis, dual rear wheels were necessary to support a 350-gallon water tank plus a hose load. (Shaun P. Ryan)

This classic 1939 model used by Freeport, Maine, was among American LaFrance's early experiments with closed cab rigs. (Dick Bartlett)

American LaFrance's "Invader" 500 Series pumper was considered a radical, stylish design during the 1930s. The 500 Series was built from 1938 to 1942. This 1938 500 was powered by a V-12 motor and was rated at 500-GPM. The attractive pumper served Monhagen Hose & Salvage Company No. 1 of Middletown, New York. (Michael Martinelli)

The Americam LaFrance Type 75 was the standard workhorse of its day. This 1920s, 750-GPM Type 75 ran as Engine 18 of the Memphis, Tennessee, Fire Department. A squirrel tail suction hose wraps around the front portion of the rig, and the trademark ALF bell with eagle is mounted to the rear grab rail. (Author's Collection)

Proving its resourcefulness, in 1930, the shop of the Milwaukee Fire Department built this 1,000-GPM pumper from the ground up. Having a Four Wheel Drive (FWD) chassis, the rig was labeled an "MFD-FWD." From 1926 to 1931, Milwaukee's shop turned out 12 pumpers, 6 ladder trucks, and 3 hose wagons. (Chuck Madderom)

Engine 1 of the Menomonee Falls, Wisconsin, Fire Department is a beautifully restored 1930 Seagrave Suburbanite 6, rated at 350-GPM with a 100-gallon water tank. Gas masks were kept in a wooden container on the running board, and a booster hose line was stored in a basket atop the pumper. (Chuck Madderom)

The Fire Department of New York, in 1938, ordered 22 Ahrens Fox Model HT 1,000-GPM pumpers, two of which were actually purchased by the 1939 World's Fair. They were turned over to the FDNY in 1941. This pumper was the second of the pair delivered, which served the New York World's Fair Fire Department as Engine Company 2. Its features included a Hercules engine, Invincible deck pipe, rear step windshield, and split-front windshield with independently-opening panels. Fifty Ahrens Foxes served the FDNY from 1915 to 1972. (Author's Collection)

The Simsbury, Connecticut, Volunteer Fire Company acquired this 1935 International/American LaFrance C Model when it assumed responsibility for the town's fire protection in 1944. The 500-GPM pumper served more than 30 years and was restored in 1991. (Cliff Williams, Simsbury Volunteer Fire Company, Simsbury, Connecticut)

An early example of custom apparatus built onto a commercial chassis was this 1930 Dodge Brothers/Van Pelt 350-GPM pumper with a 300-gallon tank. A front-mounted pump allowed more room in the truck body for hose and a large water tank. While not as glamorous as trucks built by big-name competitors, its simple design proved highly functional. This rig belonged to the Columbia Volunteer Fire Department of Tuolumne, California. (Shaun P. Ryan)

Tank-pumper combinations became popular for addressing the problem of inadequate water resources. The City of Milwaukee was fortunate to have acquired this rare, 1935 Mack Type 95 through annexation of the Town of Lake. The 750-GPM pumper/2,500-gallon tanker was not only believed to be one-of-a-kind, but also was the largest fire engine in the world in its day. (Gerrit Madderom)

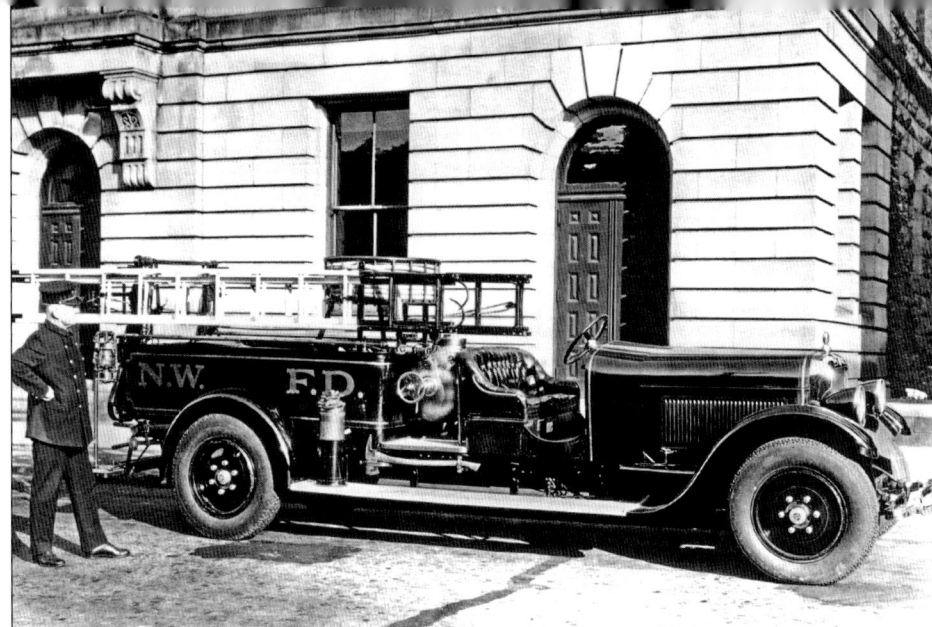

This vehicle is a 1921 American LaFrance Type 75 Extra 2 750-GPM pumper of the Grand Rapids, Michigan, Fire Department. Departments often fabricated windshields for their apparatus. (Robert J. Barber)

Its use as a fire truck did not diminish the classic lines of Studebaker apparatus, as evidenced by this 1929 pumper. (Author's Collection)

Mack's Type 21 model was the most powerful of its early B Series fire trucks. In 1936, the FDNY received 21 Type 21s, which were powered by 225 HP Hercules engines, which gave a 1,000-GPM pump capability. The following year, 19 additional Type 21s would go to New York, however, they featured closed cabs and rear windshields. Upon delivery to the department shop, this Type 21 would have the company number applied forward of the "FDNY" lettering. "Subway straps" above the rear step were a common feature on FDNY pumpers. (Author's Collection)

One of New York's rarest pieces of fire equipment was Engine 324's pumper, which was experimentally built by the FDNY shops in 1939. New York's mayor believed that shop-built apparatus would be cheaper than purchasing new rigs. In the end, the total cost was three times more than a factory-built pumper. This unique rig combined a commercial chassis, a 1,000-GPM ALF pump, a Hercules engine, Westinghouse-Bendix air brakes, and bodywork designed by FDNY and fabricated by Ward LaFrance. After assignment to Engine 324, which was organized in 1940, the rig was displayed at the 1939 World's Fair. (Author's Collection)

The Ahrens Fox is considered the classic of all fire apparatus. Among the largest of all "Foxes" ever produced is this 1,300-GPM Model PS-4, which served Salina, Kansas, for 40 years, beginning in 1929. The Fox was known for its trademark front-mounted pump with a large spherical tank. Usually chromed, the tank absorbed pressure variances as the pistons operated. (Author's Collection)

Mack introduced its E Series fire apparatus in 1937 and produced 13 types within the series until 1950. The largest and most powerful Mack E was the Type 21, which was powered by a 225 HP Hercules engine. Pumps ranged in size from 1,000 to 1,500-GPM. Although popular with large cities, only 38 Type 21s were built. This Type 21 sedan cab pumper is seen prior to delivery. Its distinctive compound-curve rear body was discontinued on later models. (Author's Collection)

New 1939 Mack pumpers for Joplin, Missouri, are displayed at department headquarters. Although outwardly identical, three are Type 75s, rated at 750-GPM, while the fourth is a Type 50, rated at 500-GPM. Enclosed cabs accommodated three people. Besides squirrel tail suctions, all were equipped with side cab-mounted spotlights and bells. (Author's Collection)

Engine Company 7 of the Kansas City, Missouri, Fire Department used this 1930s Aherns Fox pumper. Spotlights were a common feature of apparatus, and they usually were mounted at the front or rear of cabs. (Gus Johnson Collection)

The Peter Pirsch & Sons Company of Kenosha, Wisconsin, built this front-pump unit on a Ford chassis in 1929 for the Bowling Green, Kentucky, Fire Department. (Author's Collection)

Engine 4 of the Portland, Oregon, Fire Department used this 1,000-GPM 1928 Mack. (Rick Howard Collection)

Seagrave sedan pumpers became popular with the Detroit, Michigan, Fire Department. Engine 31 was a 1938 Seagrave, rated at 1,000-GPM. Noteworthy is its "Sweetheart" grille, pre-connected front-mounted suction hose, and Roto-Ray revolving warning light above its split windshield. (Neil McCarten)

Fageol was a pioneer truck manufacturer that built trucks for the western market. Fageol Truck and Coach Company of Oakland, California, operated from 1916 to 1938. Their sturdy and dependable trucks are easily recognized by their distinctive radiator shells and louvered hood vents. Overcome by financial woes caused by the Great Depression, the business was sold to T.A. Peterman in 1939. Peterman, who made his fortune in the logging industry using Fageol trucks, built his own line of trucks and named them "Peterbilt." (Rick Howard Collection)

Although this 1934 Ford pumper never officially served the Omaha, Nebraska, Fire Department as a frontline unit, it has a long history with the department. After serving the Waverly and Roca, Nebraska, Fire Departments, the rig, in 1971, underwent restoration by Omaha Fire Department members and the Pioneer Hook & Ladder Association. The rig features a front-mounted pump and running boards fashioned from Omaha's Creighton University St. John's Church. In 1978, the pumper's original flathead V-8 engine was replaced by a Ford 289 cu. in. V-8, and an automatic transmission was installed. Today, as a parade rig with the Omaha Fire Department, the engine is called Engine Co. 3/4, which indicates its status outside of frontline service. (Stan Shearer)

This view shows the hose bed and rear step arrangement of the Mack E Series Type 21 sedan cab pumper. (Author's Collection)

1940s

The boom in fire apparatus design and production that carried over into the new decade changed dramatically when the nation became immersed in the Second World War. Manufacturing priorities shifted and caused new fire apparatus deliveries to dwindle. Deliveries occurred only if a vehicle was earmarked for war-production facility protection. Metal necessary for the war machine became a precious commodity, and fire departments learned to make do with the equipment they had. Equipment and manpower shortages caused by the war brought assistance from the Federal Government, which supplied auxiliary fire-fighting equipment and trained thousands of civilians as auxiliary firefighters. Thousands of trailers mounting fire-fighting equipment were delivered by train to major cities across the country, along with skid-mounted kits incorporating pumps that could mount to standard truck frames.

The proliferation of aircraft produced for World War II spawned the development of airfield "crash" apparatus, which were categorized first as "Crash Fire Rescue" (CFR) and later as "Airport Rescue and Fire-Fighting (ARFF) vehicles. As a result, ARFF units were established at civilian airports. The use of two-way radios during the war was also adopted by civilian fire departments.

During the post-war period, fire departments began updating their apparatus fleets and often took advantage of the availability of lower-priced surplus military vehicles.

Despite restraints placed on new apparatus purchases during World War Two, there was no shortage of other truck types. Milwaukee maintained its fleet of fire apparatus by converting beer trucks into fire trucks, such as this Federal, which was made available by Miller Brewing Company. By agreement, the cabs of these vehicles were not altered, however, they were fit with fire engine bodies complete with skid-mounted pumps and hose beds, along with a variety of firefighting equipment. (Author's Collection)

To compensate for equipment and manpower shortages caused by World War Two, thousands of civilians were trained as auxiliary firefighters, and hundreds of pump trailers were delivered to cities by the U.S. government. These well-equipped trailers; along with ladders, hoses, and helmets; usually arrived by rail. (Author's Collection)

Reverting commercially owned trucks to their original configuration was simply a matter of removing firefighting equipment from the truck body platform. (Author's Collection)

Like larger trucks converted into fire apparatus, smaller vehicles pressed into fire service were given skid-mounted pumps while other equipment was added wherever room allowed. (Author's Collection)

As part of the government-funded auxiliary civilian defense program, this early 1930s milk truck saw duty as a fire truck during the war. (Author's Collection)

Since the early 1900s, the Heil Company of Milwaukee, Wisconsin, was known for its welded truck bodies. Heil built this custom 800-gallon tank body on a 1948 Ford chassis for the Adelphi, New Jersey, Fire Company. Equipment compartments flanked the body, and a 200-GPM pump was housed in the rear section. (Author's Collection)

Labeled "Fire Combat Unit No. 1," this early 1940s Heil/Mack unit appears to have been based on Mack's Type 25, which used a 100 to 200-GPM pump. The Type 25 was one of 15 designs among Mack's E Series, built from 1937 to 1950. The siren mounted near the driver's door was standard on the 206 Type 25s built. Newington, New Hampshire, ordered this unit with a specially built tank body surrounded by equipment cabinets and racks. (Author's Collection)

This GMC General is viewed from the right side. Pumpers built onto commercial chassis were common during WWII. The arrangement of hard suction hose on the truck's left side, with ladders on the opposite side, became a standard in the fire service. (Janine Kozak)

Buffalo Fire Appliance Corporation built this 500-GPM pumper on a 1942 GMC General for the Avon, Colorado, Fire Department of Eagle River Fire Protection District. The pumper was originally purchased for the New Jersey Zinc Mine at Gilman, Colorado. The number of hard suction hoses indicates a long reach to the pumper's water source. (Janine Kozak)

During the late 1940s and early 1950s, White built various types of trucks in its "Super Power" series. The power from this pumper was derived from a flathead 6 motor with a five-speed transmission. This White was Engine 3 of the Merrimac, Wisconsin, Volunteer Fire Department. (Author's Collection)

"Combination 1" of the Portsmouth, New Hampshire, Fire Department used this 1941 Seagrave, which featured the firm's "waterfall" grille. (Dick Bartlett)

Beginning in the period prior to World War Two, Holibird built fire trucks for the U.S. Army. This 1940 Holibird/U.S. Army pumper was acquired by the Milwaukee Fire Department. Having a pump rated at 750-GPM with a 100-gallon water tank, and powered by a Waukesha six-cylinder engine, Code 343 served two frontline and two reserve engine companies before being sold in 1965. (Gerrit Madderom)

Despite their popularity, having an American LaFrance apparatus was rare for the Milwaukee Fire Department, mainly since other major manufacturers were in proximity to the city. Featuring its popular cab-forward design, this 1948 700 Series was acquired by Milwaukee through annexation of the Town of Granville in 1956. To compensate for Granville's lack of hydrants, this 750-GPM pumper carried 600 gallons of water. This classic truck remained on the MFD inventory until 1968. (Gerrit Madderom)

American LaFrance pioneered the cab-forward design with its 700 Series, which was designed in 1945, with production beginning in 1947. This 700 was rated at 750-GPM and was powered by a Model G V-12 engine with a four-speed transmission. Marion, Ohio, used this interesting two-tone gray color scheme. The signature American LaFrance bell on the front bumper is balanced by a Roto-Ray warning light. Unusual on the 700 series was the pump panel that was located on the right side of the fire truck. (Author's Collection)

Pahrump Valley, Nevada, Fire and Rescue ran this 1946 Federal with a General Detroit body. Civil Defense emblems were worn by the pumper, which had a 750-GPM Waterous pump and a 500-gallon water tank. The unit was powered by a Continental in-line, six-cylinder engine. (Chuck Madderom)

The Marmon-Herrington Company began as the Nordyke and Marmon Machine Company, which made flour mill machinery. The firm built autos until the Great Depression, when it was joined by Arthur Herrington. A line of military trucks led to the Marmon-Herrington Ford, like this late 1930s crash truck. Among this unit's extinguishing agents are high pressure fog and chemical systems (mounted forward) and carbon dioxide, soda-acid, and pyrene extinguishers on the tailboard. Two, quick-attack hose lines are pre-connected at the unit's rear, with the hose stored in wells. The firefighter wears the fire entry suit style. (Author's Collection)

Montgomery County, Pennsylvania's, Plymouth Fire Company used this 1947 Ford/Maxim with a 500-GPM Hale pump until 1979. The company continues to use dark blue apparatus. The fire extinguisher on the running board was painted the truck color. (Author's Collection)

The 1940s saw widespread use of pumper/tanker combinations, often built on surplus military vehicle chassis. The Redlands, California, Fire Department operated this 1942 Reo 6 x 6 as "Tanker 1." During its conversion, its military canvas cab cover was replaced with a more durable solid cover. Tanker 1 carried 1,000 gallons of water and had a 500-GPM pump. (Chuck Madderom)

Typical of fire apparatus built with minimal chrome during World War Two is this 1943 Pirsch of the Milwaukee Fire Department. This 1,250-GPM, closed-cab pumper served as a frontline engine for 21 years before being placed in reserve status. Black or brown roll-down grille covers were common on apparatus in areas that endured harsh winters. Covering the grille preserved engine heat. (Chuck Madderom)

Equipped with searchlights on its top deck, this 1948 American LaFrance served as Engine 27 of the Dallas, Texas, Fire Department. (Author's Collection)

Tandem rear axles, dual booster hose reels, and front-mounted pump leave little doubt about this rig's use as a tanker. Van Pelt built the stout body on a 1943 Diamond T for the Waterford-Hickman Fire Protection District of Stanislaus County, California. (Shaun P. Ryan)

Although this pumper was built around a 1948 Chevrolet, the builder omitted cab doors, in keeping with a style more common on older rigs. Front-mounted pumps on commercial chassis seldom were rated more than 200-GPM. (Shaun P. Ryan)

This 1941 Ford belongs to the Millbury, Massachusetts, Fire Department. Above the larger inlet gate on its front-mounted pump are two discharge gates. (Dick Bartlett)

This 1,000-GPM, 1946 Mack was used by the FDNY for funerals of firefighters killed in the World Trade Center attack. (John A. Calderone/*Fire Apparatus Journal*)

Acquired from Milwaukee's annexation with Granville in 1956, this 1948 FWD, with a 750-GPM Waterous pump and a 500-gallon tank, served as a reserve engine. (Chuck Madderom)

Pumpers often were the basis for conversion to specialized apparatus, while maintaining their pump capability. "Chemical 1" was a 1949 Mack converted to a 750-GPM foam unit for the Reading, Pennsylvania, Fire Department. Foam concentrate in the white containers was poured into the square top tank to be proportioned with water for discharge through the deck pipe or hand lines. (John A. Calderone/*Fire Apparatus Journal*)

Carson City, Nevada, used this sturdy 1948 Mack E Series Type 505 as a house wagon. Mack built 160 Type 505s, which were powered by Mack 415 or 510 cu. in. engines. (Shaun P. Ryan)

American LaFrance built this pumper on a 1947 Ford, which protected the South Portland, Maine, Shipyard. (Dick Bartlett)

In 1948, Mack, along with other fire apparatus manufacturers, produced record numbers of rigs to make up for limited production during World War Two. Mack's L Series production spanned from 1940 to 1954, however, the post-war era signaled a return of chrome and gold leaf trim. Among post-war Macks ordered by the FDNY, this 1948 pumper was assigned to Engine 38 in the Bronx. It was powered by a Mack Thermodyne engine driving a Hale centrifugal pump. Cones behind the rear fender held nozzles. (Mack Trucks)

During the 1960s, when greater emphasis was placed on breathing equipment, Seattle engines were given cabinets for self-contained breathing apparatus (SCBA), with the hard suction hoses relocated above them. Seattle's Engine 16 was a 1949 Kenworth rated at 1,500-GPM. (Rich Schneider)

Even the circus got in on the act during the trend when private organizations formed their own firefighting units. This pickup truck, modified with a booster tank and a hose reel, protected the Clyde Beatty Circus at Los Angeles in 1947. (Author's Collection)

When the truck industry began building larger and more powerful trucks during the 1940s, the largest vehicles operated by fire departments often were tankers. When the City of Milwaukee annexed the Town of Lake in 1954, it acquired this massive 1948 Autocar tank truck. The four-wheel drive unit had a 500-GPM pump and carried 3,200 gallons of water. Such large amounts of water gave firefighters an edge in fighting fires in rural areas without hydrants. (Chuck Madderom)

Typical of big city fire apparatus during World War Two and in view of precious metal conservation, costs, a subdued finish with very little chrome is found on this 1944 Mack Type 21 of the Portland, Oregon, Fire Department. Eighty Type 21s were built, whose pumping capacity ranged from 1,000 to 2,000-GPM. This engine was rated at 1,500-GPM. (Rick Howard Collection)

Except for its headlight trim, the absence of chrome on this M Series Studebaker dates its construction to the war years. The tanker/pumper featured an elliptical-shaped tank, around which was built a booster reel, a hose rack, and a rack for wooden ladders. Hard suction hose is carried atop the tank body. Studebaker ended car and truck production in 1964. The Erma Volunteer Fire Company protects Lower Township, New Jersey. (Author's Collection)

Milwaukee's Engine 31 illustrates the return to chrome after the war. The 1948 Mack's 750-GPM pump rating is easily identified by three discharge gauges on the pump panel, each representing 250-GPM. Being Milwaukee's last frontline engine company without a water tank meant that the pumper had to rely on a hydrant for every fire. After 1950, Mack Trucks replaced the chrome grille with red paint for greater contrast to the brilliant chrome radiator shell. (Gerrit Madderom)

Portland, Oregon, operated this hose and turret wagon/boat tender, which was built by Wentworth & Irwin on a 1948 Kenworth. The imposing rig had no pump, but it carried 2,400 feet of 3 ½-inch hose. Connected to fireboats, this hose supplied large volumes of water. The unit was retired in 1976. (Rick Howard Collection)

1950s

Fire apparatus not only was made larger during the fifties, but more specialized apparatus also was being built from surplus military jeeps converted for use in firefighting and heavy rescue, using both commercial and custom chassis. Apparatus manufacturers enjoyed this boom in production, especially with the expansion of suburban communities. In that era of the Cold War, the governmental Civil Defense (CD) network, which prepared for disasters including nuclear attack, naturally was integrated with fire departments. Civil Defense funding became available for equipment purchases, and CD apparatus was kept in firehouses, many of which had air raid sirens installed on their roofs. The emphasis placed on preparedness was evident by the FDNY's acceptance of 30 Ward LaFrance pumpers to mesh with the city's "Atomic Bomb Plan."

During this period, American LaFrance was considered the leader in apparatus production. The firm's popular 700 series cab-forward design, powered by a V-12 engine, not only was in great demand, it was adopted by competitors. In its quest to stay ahead of the competition, American LaFrance introduced its 800 series in 1956 and its 900 series in 1958. Ahrens Fox introduced its cab-forward design, which was inherited by Mack Trucks. Mack began work on automatic transmissions for fire apparatus, although they would not become popular until years later.

On the opposite coast, Crown, which was known for its truck bodies and school buses, premiered its open, cab-forward Fire Coach at the beginning of the decade. Crown apparatus became a mainstay of fire departments on the West Coast and in Hawaii.

Brass water hand-pump extinguishers, nozzles, and fittings represent an era when appliances were kept on apparatus sideboards in highly polished condition. (Author)

Brush breakers were common during the 1950s in northeastern states. This 1957 International Harvester/Thiebolt served the Falmouth, Massachusetts, area. (Dick Bartlett)

This vehicle is a 1950 Seagrave 1,000-GPM pumper, painted medium gray, of the Consolidated Fire Association of Bordentown, New Jersey. (Mark Hoeller)

At a time when industry favored the cab-forward design, Seagrave, in 1951, brought out its stylish 70th Anniversary model, with the traditional engine-forward design. Seagrave would wait until the end of the decade to feature its line of cab-forward apparatus. Maxim, which was predominant on the East Coast, introduced its cab-forward in 1959. Pirsch had a thriving business during the 1950s, with its aerial ladder trucks gaining in popularity. Another Wisconsin-based firm, the Four-Wheel-Drive (FWD) Corporation, joined the cab-forward craze in 1959.

Although not recognized as custom or commercial fire apparatus, Ford's C series cab-forward, on which the cab tilted forward to provide access to the engine, is worthy of mention. The boxy, flat-front cab, which first appeared in 1957, was used for fire apparatus more than any other commercial chassis.

Apparatus innovations during the 1950s included the domed rotating beacon, which was seen mounted atop cab roofs or windshields, along with air horns, which were a spin-off of air brakes, then finding their way onto fire apparatus.

Tank No. 2 of the Goodwins Mills, Maine, Fire Department was a 1953 Chevrolet with a brush guard. (Dick Bartlett)

The six gauges located at the top of this 1952 FWD identify this vehicle as a 1,250-GPM pumper. As compressed-air breathing equipment became more important during the 1960s, compartments for breathing apparatus replaced hard suction hoses, which were relocated to the opposite side of the fire truck body. Amber fog lamps have been added to the front bumper of this vehicle, which served as Milwaukee's Engine 14. (Chuck Madderom)

New York's Engine Co. 222, a 1959 Mack C85F, represented Mack's cab-forward design. Mack Trucks, Inc. used this cab design to compete with the popular American LaFrance type. A total of 308, 750-GPM C85s were built, along with 605, 1,000-GPM C95 and 129, 1,250-GPM C125 versions. Although two jump seats were provided, Engine 222's firefighters are seen here riding on the tailboard. (www.fireapparatusphotography.com)

Mack's Model 85 was one of six fire truck types of the L Series, made from 1940 to 1954. More than 700 Type 85s were built and were all powered by six-cylinder, Mack Thermodyne 707 cu. in., 225 HP gasoline engines. The L Series was said to be Mack's crowning achievement in trucks. The image of power and ruggedness, this 1952 85L operated with the Winchester, Virginia, Friendship Fire Company. Like all 85Ls, its pump size was 750-GPM, with a 300-gallon booster tank. On Model 85s prior to 1950, grille louvers were chrome. A Federal Solar Ray light is mounted atop the center windshield, and bumper lights were added. Most firefighters found the Mack 85L one of the best handling apparatus. (Author's Collection)

The Portland, Oregon, Fire Department displayed its new line of Maxim and Kenworth pumpers in 1951. The exception is an FWD, fourth from right. All featured bodies were built by Kenworth & Irwin, Inc. of Portland. (Rick Howard Collection)

In keeping with the Cold War mindset of the 1950s, federally-funded civil defense equipment was incorporated into fire departments of major cities. Finished in the CD blue and white livery, this 1,000-gallon tanker was assigned to Chicago's O'Hare Field. The unit, called "CFR 636," was built on a 1952 Studebaker 6 x 6 military chassis. (Bill Friedrich)

This vehicle started life as a U.S. Navy FNN-1 crash truck, which was built by John Bean on a 1942 International, 1 ½-ton, 4 x 4 chassis. In keeping with Bean's specialization of water fog, the truck was equipped with a 435-gallon water tank, a high-pressure engine-driven pump, hose reels, and fog nozzles. During the 1950s, Maxim updated the truck for the Sanford, Maine, Fire Department. (Author's Collection)

Maxim Motor Company of Middleboro, Massachusetts, began building fire apparatus in 1914 and became well known in the industry. Besides custom apparatus, Maxim built vehicles on commercial chassis such as this 1950s Dodge pumper for Packanack Lake, Wayne Township, New Jersey. (Maxim)

Built during the 1950s, this American LaFrance 700 Series pumper served the Center Line, Michigan, Public Safety Department. Hard suction hose was carried atop the overhead ladder racks. ALF's popular cab-forward design was adopted by other major fire truck builders. (Mark Hoeller)

The Milwaukee Fire Department labeled this 1955 FWD an engine/tank since it featured both a 500-GPM Waterous pump and a 1,500-gallon water tank. Powered by a six-cylinder, International Harvester motor, the rig served intermittently as a front-line engine and tank company. The pump was recessed into the slab-sided body, which featured a top-mounted ladder rack. (Gerrit Madderom)

Large industrial concerns formed their own fire departments to attack fires prior to the arrival of city equipment. After two serious fires, Miller Brewing Company formed its own department in 1947. Miller's fire department purchased this new 1950 Ford, which the John Bean Company built into a front-line pumper. Called Fire Engine 34, the unit was equipped with a 100-GPM front-mounted pump, two 300-foot high-pressure hose reels, and extension ladders. After the pumper was rebuilt in 1955, it served double duty as a crash unit at car races held at Wisconsin's State Fair Park. (Miller Brewing Company)

New York's Engine 157, a 1958 Mack C85F, was modified with a protective cover for firefighters riding on the tailboard, along with a box enclosure for Scott air pack breathing apparatus, just forward of the hose reel. The one-inch hose reel, which was positioned mid-ship for tailboard operation, was among features the FDNY specified for its 1958 order of 52 C85s. Other items included single headlights, a single front spotlight, and a roof-mounted deluge gun. (www.fireapparatusphotography.com)

To help cope with Wisconsin winters, road sanders immediately forward of the dual rear wheels were often seen on Milwaukee Fire Department Pirsch apparatus. Sand kept in the container atop the running board was manually released through the dual pipe just forward of the wheels to increase traction on icy roads. The winter grille cover is in place to retain engine heat. A chrome bell and Federal Q siren mounted on the front bumper complete the good looks of this pumper. (Chuck Madderom)

"CFR 653" of the Chicago Fire Department was a 1957 Ford Cardox unit that served the city's O'Hare International Airport. The unit carried 4,000 pounds of carbon dioxide, which was expelled through a hydraulically-controlled arm. A ground sweep nozzle is visible below the front bumper. (Bill Friedrich)

The California sun was not kind to the red finish of this 1957 Seagrave of the Hayward, Alameda Fire Department. Split "Vee" windshields and vertically arranged pump panels were typical of Seagraves of that era. Fire departments often added post-mounted rotating beacons behind the cabs of their open-cab apparatus. (Shaun P. Ryan)

Engine 6 of the Manchester, New Hampshire, Fire Department ran this 1953 Mack. (Dick Bartlett)

Carbon dioxide units, such as this 1958 O-6 White/Cardox, were common at U.S. Air Force bases during the 1950s. (Author's Collection)

Engine 6 of Portland, Maine, used this unusual Mack/Seagrave combination, built in 1950. (Dick Bartlett)

This vehicle is a 1953 Seagrave of Cary, North Carolina. (Shaun P. Ryan)

Portland, Oregon's, Apparatus No. 85 was a 1951 Kenworth. (Rick Howard Collection)

Ahrens Fox built this 500-GPM unit on a 1954 GMC chassis, and it served the Crescent Springs, Kentucky, Volunteer Fire Department. (Bruce Neal/The Antique Fire Brigade Collection)

The Chattanooga, Tennessee, Fire Department used this sturdy 1957 Ford, which had its exposed pump mounted forward of the hose/tank body. (Author's Collection)

Wearing Chicago's trademark black over red scheme, this FWD pumper was built during the 1950s. (Author's Collection)

Although Engine 257 was delivered as an open cab, the FDNY shops built covers over the cab and tailboard to protect firefighters during civil disturbances and bad weather. Engine 257 was a 1953 Ward LaFrance 85T, with a pump rated at 750-GPM. (www.fireapparatusphotography.com)

Black paint, finished with gold leaf and chrome detail, makes this 1956 Mack B95F an eye-pleaser. In service as Engine 4 of the Mount Horeb, Wisconsin, Fire Department, this well-maintained classic features a 1,000-GPM pump, along with a 500-gallon water tank. Dual deck pipes gave the rig extra hitting power. (Don Feipel)

This 1954 Ahrens Fox 750-GPM unit may have served as a quad unit with the Silverton, Ohio, Fire Department. (Bruce Neal/The Antique Fire Brigade Collection)

Mack's B Model fire trucks were known for their rugged good looks. This B Model offered a wide variety of style and equipment options. Among the eight B model types, the Type 21 was the powerhouse, with its Hall-Scott engine of 300-plus horsepower. Only nine units were built, the first of which went to Memphis, Tennessee. Besides its unique sedan cab, Engine 25 featured a 1,500-GPM pumping capacity and a pre-connected flex suction hose, which wrapped around the forward half of the truck. (Author's Collection)

Detroit's Engine 26 was a 1950 Seagrave rated at 1,000-GPM, and it featured the "waterfall grille," which was used by Seagrave from 1937 to 1951. (Author's Collection)

One of six ordered by Seattle, this 1958 Mack B Model 1,500-GPM pumper was built by the George Heiser Body Company, which specialized in fire apparatus built on Kenworth chassis, mostly for cities in the Pacific Northwest. Thin white rods with red tips attached to the ends of the front bumpers gave the driver a reference point beyond the Mack's massive hood and fenders. (Rich Schneider)

The Neep Company built this 1,500-GPM pumper on a Kenworth chassis for the Seattle Fire Department. Steel diamond tread, of which running boards were usually fabricated, was used as a "kick plate" below the cab's doors. The deluge set mounted atop this pumper could have hose attached for use as a deck pipe, or it could be quickly removed for ground operation. (Rich Schneider)

This well-used Mack B85 of the Rochester Fire Department is an early example of the deviation from the all-red fire apparatus. The B85 wore Mack Trucks' trademark bulldog in three positions on the engine cover. Rotating beacons often were added to windshield frames of open cabs. As greater emphasis was placed on the use of breathing apparatus, many fire departments added cabinets for the units where hard suction hose was originally carried. Large fixed-pipe deck pipes had two hand wheels for elevating and traversing the nozzle. (Author's Collection)

Fire apparatus built by Seagrave beginning in the late 1950s featured dual headlights and more squared fenders. Seagrave sedan 1,000-GPM pumpers were popular with the Detroit Fire Department throughout the 1950s and 1960s. (Author's Collection)

Milwaukee's Engine Co. 28 illustrates the power and beauty of Mack's B95 series, long considered the favorite among Mack fire apparatus. This 1957, 1,000-GPM pumper was "Brew City's" first engine equipped with a water tank. Its 250 gallons eliminated the need for hose layouts at every fire. Although the Milwaukee Fire Department used B95 series ladder trucks, this was the department's only B95 series pumper. Like most major cities, Milwaukee assigned its apparatus shop codes. Three-digit, 300-series numbers identified engines, with this B95 coded No. 388. (Gerrit Madderom)

Fire trucks built by Howe Fire Apparatus of Anderson, Indiana, were a common sight at Army and Air Force bases, and major cities, during the 1950s. Howe's "Defender" series was built with Gramm or Duplex bodies until the firm was purchased by Grumman in 1976. They were powered by six-cylinder Continental Waukesha, or Hall-Scott engines, and had Waterous pumps and booster tanks. This 1950 Howe/Duplex Type 750 is seen at Lowry AFB, Colorado, in 1955. An aluminum, three-section, 50-foot ladder is carried on an overhead rack. (Author's Collection)

Roney built fire apparatus at Portland, Oregon, from 1953 to 1961. The Roney emblem is mounted at the center of the grille of this 1958 International/Roney, which was rated at 750-GPM, with an 800-gallon water tank. The rig belonged to Kitsap County District 15, Washington, which is now Central Kitsap Fire & Rescue. The fire department emblem is similar to that used by Seattle. (Author's Collection)

This 1954 Seagrave pumper, which ran as Engine Co. 19 of the Portland, Oregon, Fire Department became a quad with the addition of an overhead rack to accommodate ground ladders. Seagraves of this period were easily identified by unique high cab enclosures that incorporated roof windows and the Federal Q siren that recessed into its long nose. (Rick Howard Collection)

This is a U.S. Air Force, 1950 American LaFrance O-11A. Its upward sloped front end to accommodate bumper foam turrets was continued on later crash trucks. Aluminized fire entry suits reflected 90 percent of heat to protect the firefighters. (U.S. Air Force)

This 500-GPM Howe Model 620 was built on a 1956 Diamond T. Its Diamond T emblem was worn on the hood in three positions and in narrow plates below cab windows, while the Howe emblem appears on the front bumper. (Author's Collection)

Formerly of Colorado's Milliken Fire Protection District, this 1950 GMC pumper was purchased in 1989 by the Canon City prison for use in a pilot program designed to train inmates as firefighters. Five inmates became certified as firefighters, however, the program was discontinued. The pumper then became the display seen here at the Museum of Colorado Prisons. (Janine Kozak)

1960s

The number of fire apparatus manufacturers peaked during the 1960s, however, it became survival of the fittest. Many of the older, established firms found themselves in outdated facilities and burdened with high labor costs. Some would fall by the wayside while others found mergers as a means to forge onward.

Changes in fire attack tactics resulted in widespread use of cross-lay hose beds using small-diameter, pre-connected cotton-jacketed hose. Diesel power was catching on, especially in view of its lower maintenance requirements. Mack Trucks, in particular, pushed its line of diesel-powered apparatus. The electronic siren became popular and augmented or replaced mechanical sirens, although the latter were louder and would make a comeback.

The civil unrest that rocked the nation during the second half of the turbulent 1960s directly influenced the design of fire apparatus. Since urban fire departments could be confronted with violent situations, manufacturers switched to fully enclosed cabs. Large city departments outfitted apparatus with makeshift cab covers and enclosures as well as protective covers for riding positions. Such measures, which resulted in many unusual configurations, offered some degree of protection for firefighters who had become exposed targets in areas where they were once accepted as friends. Apparatus bodies, accordingly, were altered so that tools and equipment were secured out of view and inside compartments.

This image captures three generations of firefighting in Suffolk County, New York. Pictured are a 1960 Ford/Snorkel truck, a 1930s pumper, and a hand-drawn hose cart. Representing Suffolk County Air Force Base at Westhampton Beach is a Kaman H-43B "Huskie" helicopter. Carrying a fire suppression kit and onboard firefighters, the Huskie filled the Local Base Rescue role. (Jim Burns Collection)

While the big names in fire apparatus manufacturing supplied larger cities, other firms, such as Darley, catered to smaller communities and private fire brigades by offering specialized vehicles and equipment kits that could convert any commercial chassis to fire apparatus.

It was during this decade, in 1968, that the Howe Company, which took over the Oren-Roanoke and Coast Apparatus firms, introduced its top-mounted pump panel, which gave pump operators unlimited visibility and protected them from traffic. Major cities were hesitant to adopt the design, noting that it made pump operators better targets.

Another major innovation in 1968, was Snorkel Company's "Squrt" articulating boom, which incorporated a high-pressure waterway with a nozzle. The boom was a less costly means of employing a maneuverable elevated stream than elevating platforms, and it was easily retrofit to pumper-type apparatus.

Superpumper

In mid-decade, the king among pumpers made its debut with the largest fire department in the world. It was built by the powerhouse Mack trucks and was just as aptly named "The Superpumper System." The system was the brainchild of renowned ship designer William Francis Gibbs, who had entertained the concept of a "land fireboat" since the early 1900s.

Designing New York City's largest fireboat, named "Firefighter," during the 1930s inspired Gibbs to accelerate his study of a pumper so large and so powerful, it could battle immense blazes at great distances from a water source. The technology necessary to bring the concept to fruition did not become available until the 1960s, when lightweight yet powerful diesel engines and high-pressure hose were developed for naval forces.

In 1962, Gibbs teamed with Mack Trucks, which became the general manager for the construction of a superpumper and tender. The proven high-horsepower, low-weight, and dependable Napier-Deltic T18-37C diesel engine was chosen to power a DeLaval pump capable of delivering 8,800 GPM at 350 PSI, or for greater pressure, 4,400 GPM at 700 PSI. To pull these massive powerplants, a Mack cab-over-engine model F715FSTP was chosen and was powered by a Mack V8 225 HP diesel engine. An Allison semi-automatic transmission with a power-takeoff unit drove the priming pump and the air compressor for starting the pump engine. A crane at the rear of the unit supported 12-inch rigid suction hose.

Original plans called for a companion Superpumper Tender, which was of similar size and make, to incorporate four large hose reels to carry 8,000 feet of 4 ½-inch hose. Limitations in technology and materials, however, resulted in a flat hose bed to accommodate 2,000 feet of hose. To maintain the original complement of 8,000 feet of large-diameter hose, three hose tenders of pumper configuration would each carry 2,000 feet of hose. Each had a deck pipe, or monitor, capable of discharging 4,000 GPM. The trio of "satellite" tenders added even greater maneuverability and flexibility than the original Superpumper design.

The tractor of the system's main tender was strengthened to support a high-pressure monitor. The tractor, which could be uncoupled from its trailer to position the monitor, featured hydraulic outriggers to stabilize the unit and counteract the monitor's tremendous back-pressure. When fed from four 4 ½-inch hose lines, the monitor could throw a stream 600 feet.

The tender's rear axles were steerable from a rearward-facing position atop the rear of the tractor. Initially intended to facilitate retrieving hose on the planned hose reels, the position was later removed.

When signing the contract for Mack Trucks to begin work on the Superpumper system, then New York Fire Commissioner Edward Thompson said the system "would be the most powerful firefighting equipment the world has ever known." Completed during 1965, the Superpumper's primary mode of operation was to hook up to hydrants on large water mains or other major water sources. In classic pumper relay fashion, the tender then connected hose to the pumper's discharge outlets and went to the fire, laying four lines. At the fire, the lines could supply the tender's monitor or numerous smaller hose lines. When the Superpumper and tender responded together, however, the satellites were dispersed so that they could arrive first to begin hose-laying operations. Satellites could also be used to supply other units or work as tenders for fireboats. Initially, the Superpumper system responded to second-alarm fires in areas of the city notorious for fires, along with waterfronts, and areas with inadequate water supply.

Changes to the system throughout its career included not only removal of the tender's rear steering, but replacing its original fireboat-style McEntyre monitor with a Stang monitor. Due to their heavy use, the three satellite tenders were scheduled for replacement, however, due to budget woes, in 1977, they were rebuilt by Com Coach. Although rebuilding erased their distinctive Mack profile, it extended their usefulness for eight years.

The Superpumper exceeded all expectations as a pumping station while its tender and satellites proved equally effective in killing fire, relaying water, flooding buildings, and knocking down sections of buildings. Wear and budget restraints took their toll, and the system fought its last fire on 20 February 1982. It was replaced by the Maxi-Water system that already had been phased in as a water supply system, using six 2,000 GPM pumpers, each with a hose wagon.

During nearly 18 years of service, the Superpumper system responded to 2,285 alarms and worked at 918 major fires, the peak year being 1975. Fuel consumption from the Superpumper's 400-gallon capacity diesel tanks was 137 gallons per hour when its 2,400 HP pump engine operated at full throttle. The pumper's longest continuous pumping operation lasted 12 hours, 35 minutes.

Despite the ever-increasing size and pumping capacity of pumpers, it is doubtful that the Superpumper system will ever be duplicated.

(Above) Nothing better emphasizes the slogan "Built like a Mack truck" than this 1960s photo. The never-say-die truck is Engine 26, a high pressure unit of the Boston, Massachusetts, Fire Department. The Mack B model continued pumping after being hit by a falling wall during a five-alarm warehouse fire. Even more amazing is that it was driven to the repair shop. (Mack Trucks)

(Left) Taped windows and national guardsmen marked the civil unrest faced by urban fire departments during the turbulent late 1960s. Engine Company 30, a 1949 Mack stationed in Milwaukee's inner city, used temporary protective shields. (Author's Collection)

American Fire Apparatus Company at Battle Creek, Michigan, built this 750-GPM pumper on a 1960 Diamond T chassis for the Plain Township Fire Department of Stark County, Ohio. A bell and Federal Q siren were mounted on the front bumper, and a windshield offered some protection from the elements for firefighters riding the tailboard. American Fire produced fire apparatus from 1936 to 1975. (Author's Collection)

This 1968 Mack C95F 1,000-GPM pumper served as Engine 28 of the Denver Fire Department. The department has a long history of white-painted apparatus. (Patrick Campbell)

The FDNY converted this American LaFrance 900 Series pumper to a hose wagon to carry large-diameter hose for responses to La Guardia Airport. (John A. Calderone/*Fire Apparatus Journal*)

Mack built nearly 4,000 CF series. Although diesel engines commonly powered Mack CFs, Milwaukee Fire Department's CFs, like this one, were gasoline-powered. The painted shield on the front of Engine Company 5's 1968 CF-600 was soon replaced by stainless steel. Standard with the CF-600 was a 1,000-GPM pump rating and 300-gallon booster tank. (Chuck Madderom)

Milwaukee Fire Department shops fabricated makeshift enclosures to protect firefighters, who normally rode the engine's tailboard. Engine Company 9's 1949 Mack shows firefighters' gear at the ready, along with a pre-connected suction hose for quick connection to a hydrant. (Chuck Madderom)

This quint is an example of the benefit of multiple-use apparatus for small fire departments with limited budgets and manpower. Located on California's Santa Catalina Island, the eight full-time members of the Avalon Fire Department operate from one station to protect 3,400 residents in an area of 1.5 square miles. Crown Firecoach, a mainstay in California, built this unit on a 1960s Ford "Super Duty" custom cab and chassis, along with a Pitman Snorkel. In 1957, Ford introduced its square C Series medium-duty cab-over, which was used mainly for delivery work and fire apparatus bodies. The C Series was discontinued in 1990. (Joe Handelman)

American LaFrance also built airport crash rigs, such as this 1960 6 x 6 900 Series Unit, which served as the Port of Seattle's Truck No. 4 to protect Seattle-Tacoma Airport. The all-wheel drive vehicle featured a remote and hand-operated foam turret. (Author's Collection)

Short wheel-base, stocky crash vehicles were standard on Navy carriers until the arrival of the smaller P-25. This Oshkosh MB-5 is seen aboard the amphibious assault ship USS NASSAU (LPH-4). Introduced in 1968, the MB-5 helped Oshkosh Truck Corporation assert its position in the Aircraft Rescue and Firefighting (ARFF) industry. The MB-5 carried 400 gallons of water, which, when proportioned with foam concentrate, produced 5,080 gallons of foam. Chains for securing the truck to the deck hang from the front bumper. (U.S. Navy)

Engine 73 of the FDNY was a 1,000-GPM pumper built by H & H on a 1963 International Harvester chassis. A makeshift protective cover was added above the rear of the crew cab. Engine Company 73 had a 275-gallon water tank for use with its one-inch hose reel. (www.fireapparatusphotography.com)

The Milwaukee Fire Department operated both open- and closed-cab FWD pumpers. Few, however, had booster reels such as as Engine 39's 1960 model, which featured a 1,250-GPM Waterous pump and a 250-gallon water tank. (Gerrit Madderom)

Seattle ran this 1964 Kenworth 1,500-GPM pumper. The small crew cab, added in 1990, reduced its booster tank size to 210 gallons. Dual air horns on engine cowls were common on Seattle rigs. (Rich Schneider)

The Washington D.C. Fire Department used a number of Pirsch 750-GPM pumpers built on Ford F850 chassis, some of which operated as dual engine companies. This 1969 unit, in the city's colorful livery, was assigned as Engine 7. Its front-mounted suction hose was stored in an extended bumper. (Author's Collection)

This massive 1969 White 9500D/Duplex, with a 1,250-GPM pump and a 500-gallon water tank, served the Middle River Volunteer Fire Company of Baltimore County from 1969 to 1986. (Author's Collection)

Stoughton Cab & Body Company built this 750-GPM pumper on a 1960 International Harvester chassis for the Stoughton, Wisconsin, Fire Department. Stoughton began as a wagon works in 1865. (Bruce Neal/The Antique Fire Brigade Collection)

The Mack R Model 600 Series fire truck was produced from 1966 through 1990. Nearly 400 were built in eight models, all of which, except the Model 608, were diesel-powered. The St. Louis Fire Department ran this Mack R as Engine Company 23. (Mark Stampfl)

Duplex pumpers built on Howe chassis were common in the 1960s. This 1961 Duplex/Howe with a "Cincinnati" cab served the Walton, California, Fire Protection District. (Shaun P. Ryan)

The Phoenix, Arizona, Fire Department ran this 1967 open-cab, diesel-powered Seagrave, with 1,250-GPM pump and 500-gallon water tank. Breathing apparatus was mounted for quick donning by firefighters. Multi-purpose dry chemical extinguishers were mounted on the front bumper. (Chuck Madderom)

Having wrap-around windshields and crew positions, Cincinnati-style cabs were adopted by many major fire apparatus builders during the 1960s. In this decade, light bars on cab roofs began replacing rotating beacons. This immaculate 1960 Seagrave was owned by the Champaign, Illinois, Fire Department. (Shaun P. Ryan)

During the 1960s, Seagrave offered both cab-forward and engine-forward models. This 1963 750-GPM Seagrave shows off the firm's distinctive engine cowl with a recessed siren port. This was Engine 6 of the East Haven, Connecticut, Fire Department. (John A. Calderone/*Fire Apparatus Journal*)

51

More than 200 fire companies comprise the fire protection force of Westmoreland County, Pennsylvania. This 1960, 900 Series American LaFrance 1,000-GPM pumper was assigned to one of three volunteer companies of Lower Burrell in the Kinloch section. The chrome-yellow pumper featured a high-mounted ladder rack and its breathing apparatus stored in unique enclosures. (Patrick Shoop, Jr.)

The Branchville, Maryland, Volunteer Fire Company used this 1969 diesel-powered Pirsch pumper in their unit. The extended front bumper allowed not only storage for the suction hose, but also storage for a bell, Federal Q siren, and air horns. A Roto-Ray warning light is mounted to the center windshield support. (Bruce Neal/The Antique Fire Brigade Collection)

In 1989, Hahn Motors, Inc. of Hamburg, Pennsylvania, rebuilt this 1969 Mack with an extended, fully enclosed cab for the Chester Hill Hose Company of Clearfield County, Pennsylvania. The 1,250-GPM pumper was given a 1,000-gallon water tank for work in rural areas. (Patrick Shoop, Jr.)

This 1965 Pirsch Model 41B was delivered painted red to the Shorewood, Wisconsin, Fire Department, which was later incorporated into the North Shore Fire Department. Its pump was rated at 1,000-GPM, with a 300-gallon booster tank. Power was supplied by a Waukesha in-line six-cylinder gasoline engine. Pirsch emblems were attached above both headlights and both rear wheels. (Chuck Madderom)

Ward LaFrance built its "FireBrand" model between 1959 and 1962. This 1960 FireBrand served as New York's Engine Company 260. Standard operating procedure for FDNY engine companies during the 1960s had the hose reel nozzle draped over the rear grab rail for quick use by firefighters on the tailboard. (www.fireapparatusphotography.com)

With its origin easily recognizable, this pumper, like many, lived out its usefulness with a commercial company. During its heyday, beginning in 1962, this American LaFrance pumper, built on a Ford chassis, served the Valdese, North Carolina, Fire Department. (Robert Brackenhoff)

Peter Pirsch & Sons of Kenosha, Wisconsin, introduced its roomy cab-forward design during the early 1960s. In 1963, Milwaukee took delivery of two closed-cab Pirschs rated at 1,000-GPM and having 200-gallon booster tanks. (Chuck Madderom)

Although extremely popular, the C Model cab-forward was not an original Mack design. In 1956, Mack purchased the C.D. Beck Company, which had taken over production of the design from the financially troubled Ahrens Fox Company. Mack produced 1,055 C models from 1957 to 1967. Between 1962 and 1967, Milwaukee purchased ten C model apparatus, six of which were C95 1,000-GPM pumpers. The first, seen here, arrived in 1962 and went into service as Engine Company 24. Mack later offered fenders or compartments around the rear wheels, with Milwaukee opting for the latter. (Chuck Madderom)

Denver's Engine 14 was a 1969 American LaFrance 900 Series Metropolitan 1,250-GPM pumper, which was the city's first diesel-powered pumper. The wrap-around diamond tread on the lower portion of the cab was added during accident repair, which gave the appearance of a Century Series pumper. (Patrick Campbell)

The Niagara Falls, New York, Fire Department ran this 1961 American LaFrance 900 Series pumper, which was rated at 1,000-GPM, with a 300-gallon water tank. Having little bumper area, the pumper had its bulky Federal Q siren mounted forward of the windshield. American LaFrance's 900 Series was best distinguished from its successor Century Series by the absence of wrap-around metal on the cab, above the bumper. (www.fireapparatusphotography.com)

Engine Company 1 of the Appleton, Wisconsin, Fire Department shows off Pirsch's distinctive cab-forward design, which the firm introduced during the 1960s. This 1968 Pirsch was equipped with an 817 cu. in. Waukesha engine, which powered a 1,250-GPM pump. Two 200-foot booster reels were carried. Appleton's department began as a volunteer fire department in 1882. (Author's Collection)

The Superpumper's six-stage, centrifugal Napier-Deltic pump was mounted at the rear of the trailer, the deck areas of which were surrounded by 16 hose connections, equally divided as inlets and discharge outlets. The two large inlets at the rear of the unit accommodate 12-inch suction hoses, which the crane supported while drafting. (John A. Calderone/*Fire Apparatus Journal*)

The companion tender, which was of similar configuration of the Superpumper, has its original fireboat-style McEntyre monitor mounted to the tractor. The forward portion of the trailer was a walk-in compartment. (John A. Calderone/*Fire Apparatus Journal*)

The tender's tractor outriggers are visible, stowed upright and rearward of the deck ladder. The eight-inch barrel of the Stang "Intelligiant" monitor could be fit with various size tips, plus a 2,000-GPM fog nozzle. The monitor, which had a large bow to reduce turbulence, was operated by hand wheels. (John Peter Maguire)

For a brief period after delivery, the tender featured a tiller position for unlocking and steering the rear axles to facilitate hose-loading. (John A. Calderone/*Fire Apparatus Journal*)

This is the rear of the tender after the tiller position was removed. The tender's hose bed was divided to accommodate 1,000 feet of 4 ½-inch hose on each side. (John A. Calderone/*Fire Apparatus Journal*)

King among pumpers was New York's Superpumper. Pulled by a Mack commercial tractor, its massive trailer contained the diesel engine and pump rated at 8,800-GPM. Air tanks, behind the cab, provided 450 psi air pressure for starting the pump engine. The forward half of the trailer served as air intakes. (John A. Calderone/*Fire Apparatus Journal*)

Three, lower-positioned 4 ½-inch hoses supplied water to the Superpumper while the higher hose at left sent water out. The mechanical crane replaced an earlier hydraulic type. (John A. Calderone/*Fire Apparatus Journal*)

Doors provided access to the Superpumper's diesel engine and operator's pump panel. Twelve-inch, hard suction hoses were stored in recessed wells below the pump. The large cylinder at the top forward end of the trailer is an exhaust silencer. The Superpumper weighed more than 34 tons, measured 43 feet in length, and was 8 feet wide and 11 feet high. Its massive size prompted careful studies of firehouses from which it could respond and streets in New York through which it could maneuver. (Author's Collection)

Since the Superpumper's tender could not be built with large hose reels as originally designed, three Satellite tenders were added. Built on Mack C Model pumper frames, the satellites each carried 2,000 feet of 4 ½-inch hose and mounted smaller versions of the tender's Stang Intelligiant monitor. Satellites were powered by 176 HP Mack diesel engines, and their monitors could deliver 4,000 GPM. (Author's Collection)

1970s

Foremost among the changes in fire apparatus during the 1970s was Ward LaFrance Corporation's offering of fire vehicles painted lime yellow. Although traditionalists stood by the color red, a moderate percentage of vehicles in the high-visibility scheme were ordered. Competing manufacturers followed suit, with some of them reporting substantial sales of lime yellow apparatus.

Orders for diesel-powered apparatus continued to rise, with the FDNY's first-line fleet becoming all diesel-powered in 1975. Also during the decade, the number of firefighters assigned to fire units declined, and the popularity of smaller apparatus built on commercial chassis increased. More apparatus were equipped with various types of foam extinguishing agents, including Aqueous Film-Forming Foam, better known as AFFF or "Light Water."

In keeping with trends to improve crew safety, Hahn Motors, Inc., which specialized in custom apparatus, offered not only enclosed tiller cabs for its ladder trucks, but also bulletproof windows. Due to ever-present budget constraints, the rebuilding of apparatus, especially by large cities, was on the rise during the 1970s.

This 1970 Pirsch Model 41B of the Germantown, Wisconsin, Fire Department has a Waukesha six-cylinder gasoline motor powering a 1,000-GPM Hale pump. In a rural area with limited water supply, the pumper had a 750-gallon water tank. (Chuck Madderom)

Boston's Engine 17 works at a fire in October 1982. The 1976 Ward LaFrance was rated at 1,500-GPM. (Scott A. LaPrade)

The Jonestown, Indiana, Fire Department operates this 1975 Ford C/Pierce 1,000-GPM pumper painted with a Bicentennial scheme that included a Minuteman on its body. (Author's Collection)

Pumpers built by Hahn were common in the Boston Fire Department during the 1970s. Like most Boston engines during the period, this reserve engine is equipped with a front bumper rail. (Scott A. LaPrade)

Boston experimented with yellow apparatus during the 1970s. With its soft suction hose hooked to a hydrant, Engine 37's Hahn wears updated markings plus the name "Huntington Express" on the cab doors. (Scott A. LaPrade)

Airport fire apparatus was one line of specialty trucks built by Walter Motor Truck Company of Voorheesville, New York, from its inception in 1898 until 1997. "Crash Unit 2" was a 1975 Walter, which protected Portland, Oregon's, airport. This Model B-1500 was the company's mid-size crash vehicle equipped with 1,500 gallons of water and 500 gallons of foam. Dry chemical fire extinguishers are mounted to the rig's front bumper, and a litter basket is seen atop the truck body. (Author's Collection)

The Seventh District Volunteer Fire Department, Inc. of Avenue, Maryland, used this 1977 Dodge four-wheel-drive "Power Wagon" as a brush unit. Pickup trucks were a popular alternative to more expensive custom-built apparatus, especially since pickups could easily accommodate a slide-in fire pump unit in the cargo box. A popular military version of this 4 x 4 truck was called the M880 series, of which 44,000 were built by Dodge during 1976 and 1977. (Mike Wilson)

Formerly of Fresno County, California, Engine 287 of the Lindsay Fire Department is a 1975 Ford C1000 mated, in 1977, with a Van Pelt body. In addition to a 1,000-GPM pump and a 500-gallon tank, the pumper has a 250-GPM auxiliary diesel pump. (Chuck Madderom)

Engine 3 was one of two CF-600 pumpers delivered to Milwaukee in 1970 that were rebuilt during the early 1980s by Motor Truck Body Company. Its 300-gallon tank was replaced by a 400-gallon tank, and the MFD shop replaced its original Mack six-cylinder gasoline motor with a diesel engine. Modifications included an extended, fully-enclosed cab and roll-up compartment doors. (Gerrit Madderom)

Chicago's Engine 92 was a 1977 American LaFrance Century pumper, rated at 1,500-GPM with a 500-gallon tank. Large cabinets along the rear body took the place of the multiple hard suction hoses normally carried. Throughout the 1970s, Chicago operated a number of Century Series with pump ratings between 1,000 and 2,000-GPM, plus 500-gallon tanks. The rigs' standard equipment included roof-mounted air horns and an extended front bumper to accommodate a pre-connected soft suction hose. (Author's Collection)

Albert Goodrich, who was commissioner of the Chicago Fire Department from 1927 to 1931, had a nautical background, and thus, he ordered that the scheme for lighting vessels, red light on port and green light on starboard, be applied to Chicago's fire apparatus. The apparatus doors on Chicago's firehouses were similarly marked. Engine 67 was a 1973 Ward LaFrance Ambassador with a pump rated at 1,500-GPM and a 500-gallon water tank. Chicago's unique black-over-red cabs are a traditional carryover from the days when the department's first covered chief's cars had black canvas tops that could not be painted. (Author's Collection)

Milwaukee's long association with Macks included Engine Company 3's 1970 CF-600, rated at 1,000-GPM with a 300-gallon tank. The rig would serve as an example of how fire departments extended the life of their trucks. (Gerrit Madderom)

Milwaukee's first and only Ward LaFrance was also its first diesel-powered rig. This 1971 Ward LaFrance, which was rated at 1,000-GPM with a 300-gallon tank, went into service as Engine 2 to protect the downtown area. (Chuck Madderom)

The City of Denver inherited this 1978 Ward LaFrance "Patriot" with the acquisition of Lowry Air Force Base. The 750-GPM pumper is seen here after a rebuild by KME in 1990. The pumper has a 600-gallon water tank and a 55-gallon foam tank. (Patrick Campbell)

Engine 250 of Los Angeles County's City of Commerce was a 1975 Crown Model CP-100-85D. In addition to its 1,000-GPM pump, the rig was equipped with a Waterous 250-GPM power-takeoff pump and a 500-gallon water tank. Power was supplied by a Cummins NTF-295 engine with Allison automatic transmission. Crowns purchased by LA County from 1974 through 1977 were of identical specifications. (Chuck Madderom)

Engine 1 of the Chicago Fire Department was a 1972 American LaFrance 900 Series, rated at 2,000-GPM. Century models purchased later in the 1970s did not have the standard bumper seen here. In addition, models ordered later used soft suction hose in place of twin booster reels atop the truck body. (Author's Collection)

Although not as glamorous as its big city cousins, this 1975 Ford/Pierce four-wheel-drive pumper is built for rugged duty. The few markings on this brute are those of Rhode Island's Forestry Division and Smokey the Bear. Smokey the Bear was adopted by the Forest Service and the War Advertising Council in August 1944 as the enduring symbol of the prevention of forest fires. A live Smokey became part of the fire prevention campaign in 1950 when a burned bear cub that survived a fire in New Mexico's Lincoln National Forest was nursed back to health. Smokey lived at the National Zoo in Washington, D.C. until his death in 1976. He is buried at Smokey Bear Historical State Park, Capitan, New Mexico, near the forest where he was rescued. (Dick Bartlett)

This 1976 Oshkosh was assigned to NASA at Moffett Field at Santa Clara, California. (Shaun P. Ryan)

Oakland Airport Fire Rescue, Alameda County, California, used this British-built, 1975 Reynolds Boughton/Chubb, characterized by a large foam monitor over the cab. (Shaun P. Ryan)

Engine 210 of the Phoenix, Arizona, Fire Department was a 1979 Mack CF, rated at 1,500-GPM with a 500-gallon tank. (Author's Collection)

The Clymer Fire Company of Indiana City, Pennsylvania, ran this 1979 Ward/Mack, which was rebuilt in 1989 with a 1,500-GPM rating and a 500-gallon tank. (Patrick Shoop, Jr.)

Virginia Beach Fire Department's hose tender 8 (HT-8) was a 1976 Seagrave rated at 1,250-GPM. HT-8 carried two deck pipes plus a deluge set. (www.fireapparatusphotography.com)

The Worcester, Massachusetts, Fire Department used this chrome-yellow 1975 Maxim pumper. (Dick Bartlett)

The Ward LaFrance Company was largely responsible for the introduction of the color lime yellow for fire apparatus. Detroit's Engine 17 ran this 1974, 1,000-GPM Ward LaFrance. (Author's Collection)

The Oshkosh P-4 was introduced during the 1970s, which signaled the progression of the firm's larger line of crash trucks. This 1976 P-4 was assigned to the 111th Fighter Group at Willow Grove, Pennsylvania. The crash truck was rated at 1,000-GPM and carried 1,500 gallons of water plus 180 gallons of foam. (John A. Calderone/*Fire Apparatus Journal*)

Resplendent with chrome, this American LaFrance typifies the trend in modernizing older apparatus. Allegheny County, Pennsylvania, updated this 1971 American LaFrance pumper by having Kenco enlarge the truck, which gave it a 2,000-GPM pump rating and a 600-gallon tank. (Patrick Shoop, Jr.)

General Motors Corporation (GMC) trucks were one type of numerous commercial chassis built by Howe into fire apparatus. This 1970 GMC 4000/Howe was operated by the Fremont, New Hampshire, Fire Department. The Howe nameplate was worn on the front edge of the hood. Howe was purchased by Grumman in 1976. (Dick Bartlett)

Like a number of fire trucks used for television roles, this 1,000-GPM, 1977 Mack CF wore fictitious markings. Engine 11 does not exist in the FDNY, but 11 was used for the pumper's appearances in the TV series *Law and Order*. (John A. Calderone/*Fire Apparatus Journal*)

The Rockwood Fire Company of Somerset, Pennsylvania, used this 1973, 1,500-GPM American LaFrance pumper. Tandem rear axles were necessary to support a 1,500-gallon water tank. (Patrick Shoop, Jr.)

Painted dark brown, Engine 151 of the Fawn Township Volunteer Fire Department, Allegheny County, Pennsylvania, was a 1979 Mack, 1,000-GPM pumper with a 1,000-gallon tank. In 1993, the Kenco Company of Ligonier, Pennsylvania, rebuilt the rig with a fully enclosed cab. (Patrick Shoop. Jr.)

Chicago's Engine 108 is a 1974 Mack MB/Howe with a 1,250-GPM pump capacity and a 500-gallon tank. (Author's Collection)

During the 1970s, American LaFrance offered its "Pioneer" series custom trucks as a lower-price alternative to its Century Series. This 1978 Pioneer of Cambria City, Pennsylvania, was rebuilt by Keystone of Lancaster, Pennsylvania, in 1995. The 1,250-GPM pumper featured a 500-gallon tank and a fully enclosed cab extension with crew doors. (Patrick Shoop, Jr.)

Since Riverside County Fire Department has contract associations with the California Department of Forestry, its fire stations house a mix of state, city, and volunteer equipment and crews. Among them is this 1979 Crown Model CP-125, rated at 1,250-GPM and having a 500-gallon tank. Power is derived from a 370 HP Detroit diesel with Allison automatic transmission. Crown apparatus built during the late 1970s had larger cabs incorporating larger windows. Mounted atop the light bar is a 3M "Opticon" unit, which used a flashing light to signal traffic lights at intersections to turn green for emergency apparatus. A modern traffic priority system is a GPS-based unit using infrared sensors, which not only reduce collisions, but also increase response time. (Chuck Madderom)

This 1972 Mack, 1,000-GPM pumper was rebuilt by Micro, in 1996, for the Wheatland Volunteer Fire Department of Mercer City, Pennsylvania. (Patrick Shoop, Jr.)

Seattle's long line of Kenworth pumpers included this 1971 1,500-GPM unit, which served Engine 40. (Rich Schneider)

Denver's Engine 30 is a 1974 diesel-powered, 1,500-GPM Ward LaFrance "Ambassador." (Patrick Campbell)

The Sharonville, Ohio, Fire Department used this 1971 Pirsch, which featured "tunnel" lights faired into the cab roof. (Bruce Neal/The Antique Fire Brigade Collection)

In 1998, Fire Cab rebuilt this 1978 American LaFrance for the Cheswold Fire Department of Kent County, Delaware. The pumper featured side-facing light bars on its fully enclosed cab and a 1,000-gallon water tank. (Patrick Shoop, Jr.)

Tele Squrt 11 of the Denver Fire department is a 1976 American LaFrance Century Series "Duo Chief-Snorkel," with a 50-foot Tele Squrt boom and aerial ladder. The rig pumps 1,500-GPM and has a 500-gallon tank. (Patrick Campbell)

Engine 31 of the Denver Fire Department is this 1977 Seagrave, rated at 1,250-GPM with a 500-gallon tank. (Patrick Campbell)

To celebrate its 100th anniversary, American LaFrance, in 1973, introduced its Century Series pumper. Engine Company 29 of the Memphis Fire Department was located on Elvis Presley Boulevard and named its Century pumper after the famous entertainer, who called Memphis home. Seen during the 1970s, Engine 29 has a 3M Opticon unit mounted forward of the light bar on the cab roof. Succeeding pumpers assigned to Engine Co. 29, which is housed with Truck 19 to form Unit 6, continued wearing the name Elvis Presley. (Author)

Engine 30 was one of four 1972 American LaFrance 1000 Series "Metropolitans" purchased by Denver in 1972. The rig was rebuilt by Denver's shop in 1989. The diesel-powered, 1,250-GPM pumper mounted an Opticon unit at the base of its windshield. (Patrick Campbell)

Still wearing the markings of Engine 33 of the Nahant, Massachusetts, Fire Department, this 1979 Mack CF chassis was completely out of its element when it was converted to a road sander. The town of Nahant purchased two such pumpers with Federal disaster funds after a blizzard in 1978. Engines 33 and 31 served until 2005, when they were sold and converted to sanders by a local landscaping company. (Ernest Tozier)

Oshkosh Truck Corporation introduced its P-15 ARFF vehicle in 1977. The massive rig features two 495 HP Detroit diesel V-8 engines and has a 6,000-gallon water capacity. When the water is combined with foam concentrate, this truck can discharge 60,000 gallons of foam. This U.S. Navy P-15 gets a workout at NAS Roosevelt Roads, Puerto Rico. (U.S. Navy)

Wearing the traditional livery of the Baltimore City Fire Department, Engine 55's 1974 Seagrave 1,000-GPM pumper, with a 500-gallon tank, remained in service until 1993. (Author's Collection)

In 1976, Milwaukee began purchasing apparatus built on a variety of commercial chassis, usually in pairs. Engine Company 33's 1977 Pirsch/White was rated at 1,250-GPM and had a 400-gallon water tank. (Chuck Madderom)

This 1973 Century Series pumper was Milwaukee's first American LaFrance since 1961 and the department's first 1,500-GPM pumper. The unit featured a 400-gallon tank and a cab roof-mounted deluge gun. It was later assigned to the fire academy. The pre-connected, 1 ½-inch hose in the tray above the pump panel is a form of the "Mattydale Lay," which first appeared on pumpers of Mattydale, New York, during the late 1940s. (Gerrit Madderom)

Pierce built the body onto this 1978 Mack MB for the Glendale, Wisconsin, Fire Department. The rig later became Engine 25 of the North Shore Fire Department. (Author's Collection)

American LaFrance's "Pacemaker," introduced in 1972, was priced between the firm's Pioneer and Century Series apparatus. This 1973 1,000-GPM Pacemaker served the Baltimore Fire Department from 1973 until 1987. (Author's Collection)

Howe built this 750-GPM engine on a 1973 Ford C series chassis for the Grapeville, New York, Volunteer Fire Company. Its color, of course, was purple. The pumper featured a top-mounted pump panel, which gave the operator all-around visibility. (John Peter Maguire)

1980s

The 1980s saw a boom in the rebuilding of apparatus as well as a dramatic increase in their size. Larger, fully-enclosed cabs appeared, some of which could accommodate firefighters standing upright. Included in many of the rebuilds were roll-up compartment doors, which had seen extensive use on European fire apparatus.

Fire departments established Paramedic programs, and Emergency Medical Services (EMS) became more of a priority with the number of medical alarms often surpassing responses to fires. To meet the demands of this extended service, some departments had pumpers built or modified to carry emergency medical personnel and equipment.

In 1985, Emergency One introduced its "Hush" design. Having its engine located in the rear of the truck made the cab more spacious and quieter while better distributing the vehicle's weight.

This decade marked a number of changes in the corporate arrangement of apparatus manufacturers. Crown, a stalwart in the industry, built its last Fire Coach in 1985. Mack Trucks, a leader in the truck industry, announced, in 1984, that it would no longer build fire truck bodies. Mack's most produced fire apparatus chassis and cab, the CF series, often was heightened or extended rearward by companies that built bodies for Mack chassis.

Vertically and horizontally extended, fully enclosed cabs began appearing during the 1980s, allowing added room and comfort for firefighters. This 1981 Duplex was converted with the large cab by E-One in 1992. The 1,750-GPM pumper, with a 500-gallon tank, served the Manor Township Fire Company of Armstrong County, Pennsylvania. (Patrick Shoop, Jr.)

In 1987, the St. Louis Fire Department replaced its fleet of pumpers with quints having either a 50-foot Tele Squrt or a 75-foot LTI ladder. The units were KME "Fire Foxes" with Waterous 1,500-GPM pumps and 400-gallon tanks. Fifteen were ordered and all were powered by Detroit diesels. Engine 8 is a Pierce Arrow unit with a 500-gallon water tank and a 15-gallon foam tank. These units were labeled engines, however, larger tandem-axle units with 110-foot aerials ordered in 1989 were called Hook & Ladder companies. In 1979, Pierce had acquired the rights to use the Pierce Arrow name, which identified classic automobiles during the early 1900s. (Dennis J. Maag)

In 1983, Milwaukee bought Pirsch pumpers, built on Louisville/Fords, the department's first four-door cabs. Roll-up compartment doors and front-mounted suction hoses were standard on the trucks, which had 1,250-GPM pumps and 400-gallon water tanks. (Mark Hoeller)

Unique on some Western States' fire apparatus was the Barton-American Intra-Cab pump configuration. Spartan built this 1,250-GPM pumper on a Western States' chassis for Portland, Oregon. Engine 11 also featured a 750-gallon water tank. (Author's Collection)

This olive drab Oshkosh P-4 of the Air Force 437th Military Airlift Wing is seen participating in Exercise Camille at Melville Hall during May 1987. (U.S. Air Force)

Engine 74 of the Houston, Texas, Fire Department represented an unusually large municipal order. Pirsch was awarded a contract to build 16 pumpers on Spartan chassis. Pirsch filled the tall order in a nine-month period during 1982 and 1983. The pumpers were equipped with 1,500-GPM pumps and 500-gallon water tanks. They were powered by Detroit 8V-71N diesels with Allison HT-740 transmissions. (Author's Collection)

Engine 5 of the Chicago Fire Department was a 1982 Ford C8000 1,250-GPM pumper built by E-One. (Author's Collection)

Firefighters aboard a P-15 prepare to attack, with foam, a fire caused by the explosion of a 40,000-gallon JP-4 fuel storage tank in May 1986. (U.S. Air Force)

Firefighters at Alaska's Elmendorf Air Force Base, in January 1983, stand ready at foam turret positions atop their P-15 crash truck named "The Big Dipper." (U.S. Air Force)

Germantown, Wisconsin, ran this 1989 Mack MC with a Marion body. The rig was powered by a Mack E7 350 HP diesel, which drove a 1,250-GPM pump. Its tank capacity was 1,850 gallons. (Chuck Madderom)

Mack Trucks built only a few of these units for structural and airport firefighting. Atlanta's "Yellow 5" is seen here in 1987. (The Fireman)

An Oshkosh DA-1500 ARFF truck, nicknamed "The Dragon Wagon," is put through field trials in 1985. The fire department of Oshkosh, Wisconsin, operates six stations, one of which houses three airport crash trucks to cover Wittman Field, home of the Experimental Aircraft Association. The DA-1500 typically had a 1,500-gallon water tank, a 750-GPM pump, a 265-gallon foam tank, and 500 pounds of Halon. Power was supplied by a Detroit diesel engine with Allison transmission. Like many crash vehicles, the DA-1500 was designed to operate beyond the confines of paved airport surfaces. (Oshkosh Truck Corporation)

Atlanta's "Yellow 2" was an Oshkosh P-15 assigned to William B. Hartsfield International Airport in 1981. (The Fireman)

Built by Entwistle, the P-25 became the standard crash truck aboard U.S. Navy carriers during the 1980s. Designed for quick response for flight deck operations, the two-wheel-drive, diesel-powered unit features a 750-GPM pump, 60 gallons of Aqueous Film Forming Foam (AFFF), three Halon 1211 extinguishers, a 500-GPM turret, and 100 feet of 1 ½-inch hose. Nursing connections allow it to tap into the ship's pre-mixed AFFF system. Its tires are foam-filled. This P-25 is aboard the USS Dwight D. Eisenhower (CVN-69). In the background is an SH-3G Sea King helicopter. (U.S. Navy)

One of four 1987 Pierce Arrow 1,000-GPM pumpers rebuilt by Pierce between 2000 and 2003, Squad 8, of Camp Lejune, North Carolina, Fire and Emergency Services Division, carries 500 gallons of water, 100 gallons of foam, and mounts a 50-foot Tele Squrt boom. (Tom Shand)

In 1985, Amertek set up shop at Woodstock, Ontario, Canada, to fill an order for 362 crash trucks for the U.S. Army after Walter Trucks went bankrupt. Amertek went out of business in 1993 after incurring losses on U.S. Army contracts. The standard Amertek truck was the 2500L pumper, such as this example, which was among more than 30 protecting the sprawling aviation facility at Fort Rucker, Alabama. A number of Amerteks remain in service, including in Iraq. Amertek fire trucks were termed "MACI," for "Military Adaptation of a Commercial Item." Numerous compartments were incorporated into its body. (Author)

The Washington, D.C., Fire Department used a number of these Ford C8000/E-One 1,250-GPM pumpers. (Author's Collection)

Easily converted to a fire truck with a slide-in pump and hose unit, this 1988 Humvee was assigned to the Camp Roberts, California, National Guard Fire Department. (Shaun P. Ryan)

This U.S. Air Force P-8 of the 63rd Civil Engineering Squadron at Norton AFB, California, helps battle the 12,000-acre Lytle Creek brush fire at Rancho Cumonga in September 1988. (U.S. Air Force)

"Red 1" is a 1982 GMC Sierra with a 1974 Fire Control-Fire Boss body used by the Aviation Department of the Department of Public Works at Denver's International Airport. For attacking fires, the unit carries 450 pounds of dry chemical, 50 gallons of pre-mixed foam, and 200 pounds of Halon. (Patrick Campbell)

Engine 8 of the Plymouth, Massachusetts, Fire Department is a 1983 Hahn. (Dick Bartlett)

Built as a low-cost alternative to its custom pumpers, this lime yellow Seagrave pumper was assigned to NS Roosevelt Roads, Puerto Rico, in 1986. (U.S. Navy)

Not all municipal fire apparatus was operated by fire departments. This 1988 Oshkosh was used by the Port Authority of New York and New Jersey. The 2,000-GPM unit carried 2,500 gallons of water and 410 gallons of foam. (John A. Calderone/*Fire Apparatus Journal*)

Engine 5 of Sterling, Massachusetts, ran this 1983 Mack, which was modified with an extended crew cab. (Dick Bartlett)

Walter used a simple box design for this 1,000-GPM pumper, operated by the U.S. Coast Guard at Governor's Island in 1984. The unit carried 500 gallons of water and 60 gallons of foam. (John A. Calderone/*Fire Apparatus Journal*)

Extensive modifications by Interstate to this 1981 Mack gave the 1,500-GPM pumper an enlarged cab, a 500-gallon tank, and a 55-foot Tele Squrt. The Sarver Volunteer Fire Company of Buffalo Township, Pennsylvania, chose a high-visibility paint scheme of white and lime yellow. (Patrick Shoop, Jr.)

Fawn Township of Allegheny County, Pennsylvania, ran this powerhouse rig, which S&S built on a 1989 GMC chassis. The 1,000-GPM unit featured a 1,000-gallon water tank. (Patrick Shoop, Jr.)

The Saltsburg Fire Company of Indiana City, Pennsylvania, used this 1985 Peterbilt/Ward LaFrance 1,500-GPM pumper with a 75-gallon tank. (Patrick Shoop, Jr.)

Keystone modified this 1981 American LaFrance 1,250-GPM pumper for Rosedale, Pennsylvania, giving it an attractive silver-over-dark red paint scheme. Extension ladders and a litter basket were carried atop its body. (Patrick Shoop, Jr.)

Denver's "Red 4" was a 1982 Oshkosh T6 4 x 4 ARFF unit rated at 1,500-GPM. It carried 1,585 gallons of water, 205 gallons of foam, and 700 pounds of Purple K. (Patrick Campbell)

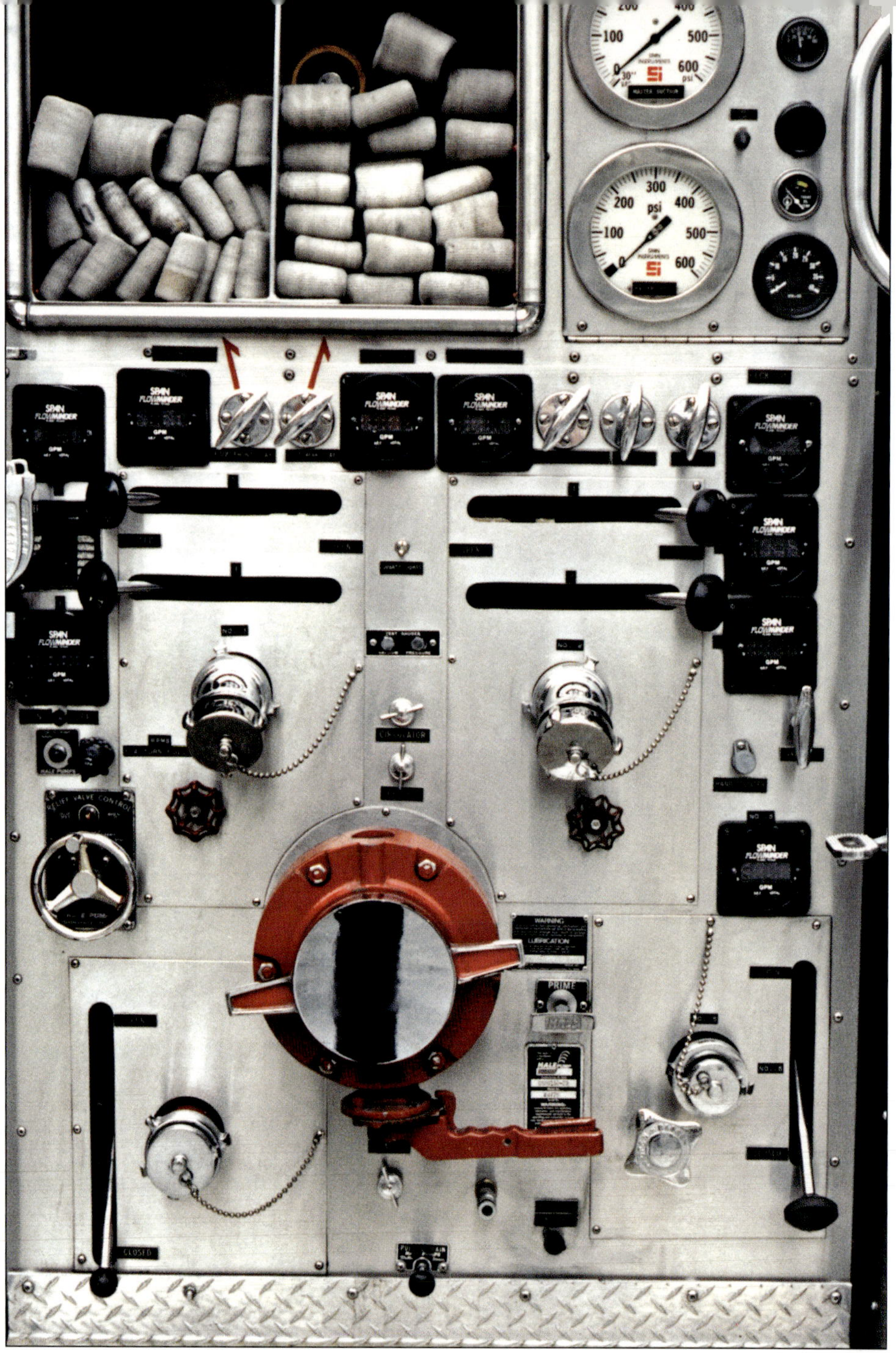

Built into the pump operator's panel of this 1,250-GPM pumper are twin quick attack hose boxes. The large red fitting at the panel's bottom is a suction inlet. (Author)

"Red 3," assigned to Denver International Airport, was a 1985 Oshkosh T2000 6 x 6 rated, which carried 1,585 gallons of water and 3,160 gallons of foam. (Patrick Campbell)

Denver's Squrt 24 was a 1989 Pemfab 1,250-GPM pumper built on a Seagrave chassis, with a 65-foot Tele Squrt built by Snorkel. (Patrick Campbell)

The largest of Denver's ARFF units was this 1987 Oshkosh DA1500 8 x 8 mated with a Rosenbauer R600NZ "Viper." Called "Mad Max," this unit was rated at 1,500 GPM and carried 1,500 gallons of water, 300 gallons of foam, and 250 pounds of Halon. (Patrick Campbell)

After Renault of France began purchasing shares of Mack Trucks, Inc., beginning in 1979, Mack began winding down fire apparatus production. In 1981, a Renault line of medium-duty pumpers became available, labeled the Mack "Mid Liner MS Model." Built for Mack by Ward 79, these aluminum-bodied vehicles featured a seven-person cab as standard equipment. This example of the California Division of Forestry shows the decidedly Renault influence. (Mark Stampfl)

Boston purchased a large number of Ford/E-One 1,250-GPM pumpers with 500-gallon tanks between 1984 and 1986. In 1987, the department switched to E-Ones. Engine 41 is housed with Ladder 14 in Boston's Allston section. (Scott A. LaPrade)

Engine 53 of the Seventh District Volunteer Fire Department of Avenue, Maryland, is a 1981 Sanford. The pumper, which is rated at 1,500-GPM, features a large 2,000-gallon tank body. The Sanford Fire Apparatus Corporation built units at its Syracuse, New York, facility from 1909 to 1937. The company specialized in the Cincinnati-style cab seen here, when it resumed fire truck production in 1969. (Mike Wilson)

St. Louis Fire Department's Reserve Engine 5 was one of 22 1977 Duplex chassis with Howe bodies, incorporating 1,250-GPM pumps with 500-gallon tanks. The unit was repainted red by the department shop in 1988 and had a large fireboat monitor installed. This was the city's only Howe engine repainted from the original white and lime green scheme. Engine 5 was special-called for its large deck pipe and remained in service until 2000. (Dennis J. Maag)

Mack Trucks, Inc. improved on its MB series by offering its line of MC custom fire trucks from 1978 to 1990. More than 700 were built. This example of the Winston-Salem, North Carolina, Fire Department featured a fully enclosed crew cab built by EEI. Although Winston-Salem switched to lime green in 1971, and then back to red in 1991, this MC pumper retained its lime green livery as a reserve engine. (Robert Brackenhoff)

This American LaFrance Century Series/Saulsbury was one of four delivered in 1982 as Superpumper System Satellites. That same year, the Superpumper System was removed from service, with the satellites forming the Maxi-Water System. This rig was later numbered 308 and served as a hose wagon for responses to New York's JFK Airport. (John A. Calderone/*Fire Apparatus Journal*)

Engine 4, which belongs to the Las Vegas Fire Department, is a 1989 Pierce "Lance" rated at 1,500-GPM. The pumper is powered by a 450 HP Detroit diesel engine, and it has a 500-gallon tank and a built-in foam eductor system. Some fire departments advertise their Class 1 Insurance Services Office (ISO) rating since it affects the fire insurance premiums of property owners in a department's area. The rating is worn on Engine 4's cab above the dice. ISO requirements differ from those of NFPA 1901 in that the ISO focuses on the fire suppression capabilities of apparatus. NFPA 1901, on the other hand, focuses on apparatus performance and safety. (Chuck Madderom)

Denver's Engine 28 was a 1984 Seagrave 1,250-GPM pumper, which the department refurbished with a jump-seat enclosure in 1999. The Denver Fire Department adopted the gold stripe and stylized letter markings in 1998. (Patrick Campbell)

Some major manufacturers; including Sutphen, Coast, and Van Pelt; used a cab roof extension over the windshield, which allowed for mounting warning lights as far forward as possible. The Columbus, Ohio, Division of Fire remained loyal to Sutphen, which is located at nearby Amlin, Ohio. Sutphen is a fourth-generation family business and the oldest continuously owned and operated fire apparatus manufacturer in America. (Mark Stampfl)

The McMurdo, Antarctica, Fire Department protects two active runways. The department is based at McMurdo, Ross Island, protects the largest geographic area on the planet, and is the southernmost fire department. Besides heaters used to keep fire equipment from freezing, an anti-freeze agent and salt-based chemicals keep extinguishing fluids usable at temperatures as low as -40 degrees Fahrenheit. Fuel fires are fought with AFFF and potassium carbonate. This late 1980s Pierce is rated at 1,000-GPM with a 750-gallon water tank. (Alan Robock)

FDNY "Ten House," located across from the World Trade Center site, is one of only two of New York City's 220 firehouses with an engine and ladder company having the same number. The other is Engine Co. and Ladder Co. 52, housed together in the Bronx. This diesel-powered 1983 Mack CF 1,000-GPM pumper served as Engine 10 until the late 1980s. Banners on its grille plate read: "Lotta Pride." The vinyl hose bed canopy and front-mounted soft suction were standard. (Author's Collection)

Oshkosh Truck Corporation introduced its 4 x 4 1,000-GPM P-19 ARFF vehicle in 1984. Tow rings mounted well below its front bumper facilitated its transport in cargo aircraft. Like most crash vehicles, Honolulu's T-3000 was painted pale yellow. (Oshkosh Truck Corp.)

Although basically a large pumping unit, Pittsburgh's Foam 1 performed multiple tasks. The 1989 Pierce was rated at 1,500-GPM. A 55-foot articulating boom could discharge its 500 gallons of water or 700 gallons of foam. (Patrick Shoop, Jr.)

Fairbanks, Alaska, purchased this Oshkosh T-3000 during the 1980s. The crash truck was rated at 1,800-GPM, and carried 3,000 gallons of water and 405 gallons of foam. (Oshkosh Truck Corp.)

1990s

In 1991, the National Fire Protection Agency (NFPA) revised its standards, which impacted fire apparatus design. The directives called for all firefighters to ride in enclosed, seated positions. In addition, Federal regulations outlining engine size, vehicle emissions, axle-loading, and anti-lock brake systems drove up the cost of fire apparatus, making the use of commercial chassis even more popular. The trend towards reducing manpower led to the production of more multi-function vehicles, which grew even larger and heavier. Onboard computers and satellite-tracking devices became common on apparatus.

During the first half of the decade, some of the big names ceased fire apparatus production. They included Crown, which closed its doors in 1991, followed by Van Pelt and Beck the following year. In 1990, Mack discontinued its production of custom fire apparatus chassis. One of the best-known names in the industry, American LaFrance, ceased fire apparatus production in 1994. Freightliner acquired limited rights to American LaFrance and built the "American LaFrance Eagle" on Freightliner chassis. The Boardman Company of Oklahoma City left the fire apparatus business in 1995. Oshkosh Truck Corporation in 1996 acquired Pierce Manufacturing, Inc. With Pierce as a subsidiary, Oshkosh became the world's leading fire apparatus producer. Compared to the hundreds of fire apparatus firms that flourished decades earlier, fewer than 40 major apparatus builders existed at the end of the 20th Century.

In 1994, the company C.R.E.S. built this 4 x 4 around an Oshkosh TB1500 for use at Denver's International Airport. "Red 6" carries 1,500 gallons of water, 400 gallons of foam, and 500 pounds of Halon, and mounts a 55-foot "Snozzle." (Patrick Campbell)

The FDNY does not have an Engine Co. 99, however, this 1997 Seagrave 1,000-GPM pumper was assigned to the department's fire academy. The markings "PFS-1" stand for "Probationary Firefighters School." Rigs assigned to the school often are used in film shoots, with Engine 99 used for the TV series *Rescue Me*. **(John A. Calderone/***Fire Apparatus Journal***)**

The Oshkosh Truck Corporation built this T series 6 x 6 crash truck for Boston's Logan International Airport. ARFF units assigned to Logan wore the motto "Committed to Excellence" on the front of their cabs. (Dick Bartlett)

Humvees modified with pumper bodies are a common site among military and civilian fire departments. This Humvee attack vehicle serves at Station 1 among 11 fire stations at the Marine Corps base at Camp Pendelton, California. (Tom Shand)

Production of the P-19 began early during the 1980s, with more than 800 built for all U.S. armed services under contract by Oshkosh, Freightliner, Rosenbauer, and JRI, Inc. Here, U.S. Army Special Forces troops provide security for an Air Force P-19 crew of the 68th Civil Engineering Squadron (CES) as it fills the rig at the Firestone plant near Roberts International Airport, Liberia. The 68th CES provided protection during Operation Assured Lift in February 1997. (U.S. Air Force)

Used as the first-response vehicle on major flight lines, the P-23 was an enlarged version of the P-19. The first batch, built by E-One from 1994 to 1996, carried 3,000 gallons of water. A second U.S. Air Force order went to Oshkosh Truck Corporation in 1999. This P-23 of the USAF 51st CES demonstrates its attack capability at Osan AB, Korea. Both front axles of the 8 x 8 unit turn independently. (U.S. Air Force)

Denver's "Mini 31" runs as the second section of Truck 31. In 1988, the firm Danko 4 Star General added an H7 Wildland Attack unit to the Humvee General to create a 250-GPM mini pumper with 200 gallons of water and 15 gallons of foam. (Patrick Campbell)

True to its name, the Greenfields Fire Company of Berks County, Pennsylvania, wears an interesting blend of green and white. The 1999 Pierce Quantum is rated at 1,750-GPM and carries 1,000 gallons of water to protect Reading Municipal Airport. (Author's Collection)

When Pierce delivered this 1,500-GPM Quantum, the North Las Vegas Fire Department was using this interesting scheme of white, gray, and red. Besides a 500-gallon water tank, Engine 56 mounted a 61-foot Pierce "Skyboom." (Chuck Madderom)

Engine 8, protecting the Pittsburgh International Airport, was built by Grumman on a Pemfab chassis. (Patrick Shoop, Jr.)

A dramatic black and white color scheme draws attention to this 1991, 1,500-GPM International Harvester/KME of the Vanport Volunteer Fire Department of Beaver County, Pennsylvania. The pump panel is in the center of the rig, behind the crew cab. (Patrick Shoop, Jr.)

Brooklyn Hose Company of Lewiston, Mifflin County, Pennsylvania, runs this 1993 E-One. The 1,500-GPM pumper boasts a 2,000-gallon water tank, requiring support from tandem rear axles. Top-mounted pump panels are common on Pennsylvania pumpers. (Patrick Shoop. Jr.)

Luverne built this 1,500-GPM pumper on a 1992 Spartan Gladiator chassis for Chicago's Engine 98. (Author's Collection)

Denver International Airport's "Red 8" is a 1992 GMC Sierra 2500SL, built by Becker using a 1982 FireTec body. The unit carries 450 pounds of dry chemical, 50 gallons of foam, and 200 pounds of Halon. (Patrick Campbell)

Baby blue Engine 2 of the Unity Volunteer Fire Department, Borough of Plum, Pennsylvania, is a 1999 Pierce 1,500-GPM pumper with 750-gallon tank. Cabs became large enough that side-facing light bars could be added. (Patrick Shoop, Jr.)

Engine 32 of the Progress Fire Department of Dauphin County, Pennsylvania, uses this impressive 1998 Seagrave, rated at 1,750-GPM. Pride shows in lettering on the cab, some of which reads, "Often Imitated, Never Duplicated," and "100% Volunteer." (Patrick Shoop, Jr.)

Denver's "Red 2" is a 1998 E-One Titan HPR built on a Federal Motors 8 x 8 chassis. Rated at 1,500-GPM, the rig carries 3,000 gallons of water, 405 gallons of foam and 550 pounds of dry chemical. (Patrick Campbell)

Denver's Squrt 10 is a 1991, 1,250-GPM Seagrave mounting a 65-foot Tele Squrt boom. Quick attack hose lines are kept on the front bumper and atop rear cabinets. (Patrick Campbell)

The Port of Portland, Oregon, Fire Department runs this 1994 Oshkosh TB-3000 2,000-GPM ARFF unit, which carries 3,000 gallons of water and 420 gallons of foam. Its roof turret is capable of discharging agents between 750 and 1,500-GPM. Power is supplied by a 575 HP Detroit diesel engine. (Jeff Holter)

This 1992 Pierce Dash was delivered to the Colorado Department of Transportation to protect the Hanging Lakes Tunnels of Glenwood Canyon. The 750-GPM pumper works with a small rescue unit and a Mack wrecker. It carries 1,000 gallons of water and 125 gallons of foam. (Dennis J. Maag)

Unusual on this 1996 E-One 1,500-GPM pumper of Junction Fire Company, Mifflin County, Pennsylvania, is a high cover over its hose body and a 1,000-gallon tank. (Patrick Shoop, Jr.)

Engine 7 of the Indianapolis, Indiana, Fire Department is a 1998 Pierce Lance II Custom 1,500-GPM pumper. Besides a 500-gallon water tank, it carries a 40-gallon foam tank. The front bumper-mounted suction enables the driver to "spot" the rig at a hydrant. (Mark Stampfl)

Pump panel fittings and a hose line match the color of this 1991 Spartan/Boardman of the Moon Township Volunteer Fire Company, Pennsylvania. The 1,250-GPM pumper features a 500-gallon tank plus a 55-foot waterway boom, with ladder attached. (Patrick Shoop, Jr.)

After the firm Delmarva extensively modified this 1990 Mack, only its front face retained the signature Mack CF look. The 1,250-GPM pumper with 750-gallon tank is operated by the Brandywine Hundred Fire Company of Bellefonte, Delaware. (Patrick Shoop, Jr.)

Fire extinguishers are fastened to the front bumper of this 1996 pumper built by Four Guys on a GMC chassis for the Elizabeth Volunteer Fire Department of Allegheny County, Pennsylvania. The 1,250-GM pumper has a 750-gallon water tank. (Patrick Shoop, Jr.)

Engine 20 of the Denver Fire Department uses this 1999 Central States/HME 1,250-GPM pumper. The high-cab engine has a green-tinted windshield. A quick-attack line is stored in a diamond tread enclosure on its front bumper. (Patrick Campbell)

21st Century

Current NFPA standards for new fire apparatus, along with manufacturer in-house engineering, have resulted in stronger, safer, more durable, and more efficient apparatus. The adoption of European cab crash-test standards has broadened the safety margin for firefighters. Cab soundproofing, exhaust-handling systems, and breathing apparatus as standard equipment on apparatus have also increased firefighter safety. Smaller, yet more eye-catching strobe warning lights are in widespread use. Aerial ladders are stronger and higher. In contrast, some experts in the fire apparatus industry express concern that some NFPA requirements, such as those concerning electronic data monitoring, are unnecessary and will drive up the cost of apparatus.

As we fall on troubled economic times, the future of fire apparatus production is uncertain. Given the pattern over the years of company terminations, acquisitions, and mergers, it seems likely that only a few large manufacturers will remain. Oshkosh/Pierce has proven itself the leading producer of fire apparatus in the 21st Century while the future of other big names hangs in the balance. Constantly faced with budget constraints, fire departments learn to do more with less, and they learn to extend the life of their equipment, thereby relying less on new purchases. Federally-funded programs that became available after 9-11 will at some point have run out of steam.

Although apparatus will appear more globally influenced and futuristic, and feature new technology, possibly including alternative fuels, they will always capture our attention and interest.

Among its fleet of Oshkosh ARFF vehicles, Tampa Fire Rescue uses this E-One Titan HPR. E-One, of Ocala, Florida, offers the Snozzle, which is a piercing nozzle designed to penetrate an aircraft fuselage and discharge foam to extinguish fire in a cabin or cargo area. This Titan carries 1,500-gallons of water, 205 gallons of foam, and 500 pounds of Halon. The unit is powered by a 665 HP Detroit diesel engine. (Robert Cotnior)

Throughout its history, the FDNY used Squad companies as supplemental manpower units, which are trained and equipped to operate as an engine or a ladder company. In June 1998, fire engine companies were reorganized as squads for response to fires and major emergencies. The "Squad 55" markings on this 2004 Ferrara demonstrator 1,000-GPM engine were applied for its use in the TV series *Third Watch*. (John A. Calderone/*Fire Apparatus Journal*)

Carnegie, Pennsylvania, runs this 2001 Ferrara 2,000-GPM pumper with a 1,000-gallon tank. (Patrick Shoop, Jr.)

Rescue Engine 93, named "Pride of the Southside," is a 2001 Pierce Dash 1,250-GPM pumper with a top-mounted pump panel for Greenwood, Indiana. (Christopher Allen)

The Plum Borough Fire Department of Renton, Pennsylvania, operates this plum-colored 2001 Pierce, rated at 1,250-GPM. The cab reads: "Only getting Better." (Patrick Shoop, Jr.)

This Oshkosh Striker ARFF vehicle serves the Tampa Fire Department. (Robert Cotnior)

This Pierce Quantum engine of the Clark County, Nevada, Fire Department features a stand-up compartment for firefighters. (Mark Stampfl)

Smith Reynolds Airport of Forsyth County, North Carolina, runs this Oshkosh Striker 1500 named "CFR 1." Strikers feature roof and bumper turrets, which are joystick-controlled. Its Waterous pump is rated at 1,950-GPM, and it carries 1,500 gallons of water and 210 gallons of foam. Located on each side of the truck is a pre-connected 1 ¾-inch, 150-foot handline with a 125-GPM nozzle. (Robert Brackenhoff)

The McMurdo, Antarctica, Fire Department uses two 2004 Canadian-built Foremost Chieftan four-track vehicles, each of which is equipped with 1,200 gallons of AFFF. (Alan Robock)

This 2004 Flex-Trac, built by Foremost for McMurdo, Antartica, is assigned to Williams Field. "Crash 3" carries 1,350 pounds of Purple K agent and 200 gallons of AFFF. (Alan Robock)

This bright yellow, with green and orange trim, 2006 Pierce pumper serves Wilkins Township, Pennsylvania. The rig can pump 1,750-GPM and carries 500 gallons of water and 30 gallons of foam. Its pump operator's panel is top-mounted. (Patrick Shoop, Jr.)

An Oshkosh P-19 of the Arizona Air National Guard is loaded aboard a C-130 of the Wyoming ANG for the Hurricane Katrina relief effort in September 2005. (U.S. Air Force)

Kern County, California, uses this 2004 Rosenbauer to protect the Inyokern Airport. Included in its array of equipment is a 1,850-GPM pump, a 1,585-gallon water tank, a 20-gallon AFFF system, Halon, 500 pounds of dry chemical, and a 10kw hydraulic generator. The unit is powered by a 600 HP Detroit diesel with a six-speed automatic transmission. (Chuck Madderom)

Marine Wing Support Squadron 271 Crash Fire Rescue from MCAS Cherry Point, North Carolina, stages with their P-19 vehicle in a makeshift shelter near the flightline at Al Asad AB, Iraq, in June 2005. (U.S. Marine Corps)

Using cockpit-like controls, Senior Airman Sayward Burns of the USAF 31st CES operates a crash truck at Aviano AB, Italy, in March 2002. The pointed nozzle of a Snozzle boom hangs in front of the windshield. (U.S. Air Force)

Using its high-reach turret boom and bumper turret, an Oshkosh Striker works out at a training fire. The Snozzle aircraft-piercing nozzle protrudes from the boom. (Oshkosh Corporation)

A U.S. Air Force firefighter of the 410th Air Expeditionary Wing in Iraq powers up the pump on a crash truck in 2003. (U.S. Air Force)

Various shades of blue have become popular with Pennsylvania fire departments. This 2001 Sutphen 2,000-GPM giant of the Broughton Fire Department was no exception. (Patrick Shoop, Jr.)

Engine 32 of the Denver Fire Department is a 2003 1,500-GPM Pierce Dash 2000 all-wheel-drive unit equipped with 750 gallons of water and 60 gallons of foam. (Patrick Campbell)

Engine Company 24 of the Milwaukee Fire Department is assigned one of five Pierce Velocity 1,500-GPM pumpers. The bus-type rear-view mirrors, extending forward of the cab, were found to be lacking and therefore, unpopular with apparatus drivers. The roll-up cabinet door above the rear wheel features a recruiting graphic. A glowing feature of the Velocity is the lowered hose bed, a welcome deviation from previous engines with significantly higher hose beds. Engine 24 wears the MFD emblem on both sides of the cab. (Chuck Madderom)

Nevada's Pahrump Valley Fire & Rescue is the proud owner of "Brush 2," a 2008 International, built by High Desert Fire Equipment. The stylish front end earned it the title "7400 SFA WorkStar Meritor" in keeping with International's 2008 practice of assigning colorful names. Unusual is the unit's 752-gallon capacity water tank. Power for the two booster reels is derived from a 300-GPM Darley pump. Brush 2's pump panel is located at the rear of its tank body. (Chuck Madderom)

Taking its name from an early period, the Vigilant Hose Company of Shippensburg, Pennsylvania, operates this cream-over-red 2005 E-One. Lettering across the top of its windshield reads, "I Wish I Was A Seagrave Engine." (Patrick Shoop, Jr.)

Black-over-pale yellow is used on this 2006 KME 2,250-GPM pumper of Laceyville, Pennsylvania's, Goodwill Fire Company. (Patrick Shoop, Jr.)

Fawn Township, Pennsylvania, uses this 2005 E-One, finished in dark brown with novel trim. Lettering on its front bumper reads, "Fireslayer." (Patrick Shoop, Jr.)

This 2001 Pierce Quantum 1,500-GPM pumper shows off the Las Vegas Fire Department's new color scheme begun that year. In addition to a 500-gallon water tank, the rig features a built-in foam system. (Chuck Madderom)

The predominantly black finish of this 2004 E-One is broken by white trim and chrome. Ample warning light systems of modern apparatus compensate for trends in low visibility paint schemes. This 2,000-GPM unit serves the Franklin Fire Company of Pennsylvania. (Patrick Shoop, Jr.)

S & S Fire Apparatus Company, which was founded in 1983, designed its Wildland Ultra XT for use in rugged terrain. The XT's 6 x 6 chassis with independent suspension at all wheels allows it to take 2,000 gallons of water, 50 gallons of foam, and six firefighters into brush fires. Power is derived from a 400 HP Cummins engine with six-speed automatic transmission. Besides having multiple discharge outlets, the XT features an Akron FireFox front turret with joystick control in the cab, and two Hannay one-inch hose reels. A ten-inch dump valve at the rear allows filling of a portable water reservoir. A Kubota 42 HP turbo diesel liquid-cooled engine powers a Darley pump capable of 500 GPM. Port Orange, Florida, Fire and Rescue acquired this Ultra XT in April 2008, just in time for the region's brush fire season. (S & S Fire Apparatus)

Despite the trend toward large cab-forward pumpers, Pierce also manufactures smaller engine-forward units such as this 2004 500-GPM pumper. Pierce built the unit on an International Harvester chassis. (Patrick Shoop, Jr.)

The Green Tree Volunteer Fire Company of Allegheny County, Pennsylvania, operates this 2005 Pierce pumper, which is rated at 2,000-GPM and has a 750-gallon tank. The beauty of its emerald green scheme is enhanced by a liberal amount of chrome and gold reflective stripes. (Patrick Shoop, Jr.)

Although the FDNY's Superpumper System was disestablished, the department retained its satellite component, which is teamed with 2,000-GPM pumpers of the Maxi-Water System. Satellite 1 is a 2002 Mack/Saulsbury tender, equipped with large-diameter hose, a Stang monitor, and foam containers. (John A. Calderone/*Fire Apparatus Journal*)

In 1998, the Kern County Fire Department of Bear Valley, California, received five rigs built on International 4800 model chassis. Having only 275 HP engines, they were found to be severely underpowered. So in 2005, they were rebuilt onto Pierce Saber chassis with 400 HP Caterpillar diesel engines. The pumpers have 500-GPM Waterous pumps, 500-gallon water tanks, and 30-gallon foam systems, plus 125 GPM auxiliary pumps. The apparatus are mounted unusually high on their frames for operations in mountainous terrain. (Chuck Madderom)

This side profile of the Model 3000 shows the compartment layout and long rectangular body design. (Oshkosh Corporation)

An Oshkosh Striker 10E demonstrates its roof, bumper, and ground spray turrets. The 10E is designed to operate at worldwide Category 10 airports that will accommodate the Airbus A380. With a centrifugal pump rated at 2,642-GPM, a 3,170-gallon water tank, and a 444-gallon foam tank, the roof turret of the 10E can throw a stream 300 feet. Foam proportioning is done electronically. (Oshkosh Corporation)

Painted "U.S. Forest Service Safety Green," this USFS Model 52 was built by Choquettes & Son Truck Body of Sparks, Nevada, on a 2001 International 4800 chassis. Codes on the truck body signify its assignment as Engine 7532 of Nevada's Kyle Canyon Station, Clark County, protecting the Humboldt-Toiyabe National Forest. The 4 x 4 truck is powered by a 220 HP International diesel engine. (Chuck Madderom)

The Oshkosh ARFF Striker series cab is designed as a cockpit with the operator seated at center and surrounded by a wrap-around dashboard and joystick controls for water/foam turrets. (Oshkosh Corporation)

On the unit's rear deck is a 50-GPM pump powered by an 18 HP engine. Forward of the 700-gallon water tank are booster reels on both sides. A foam system is also mounted into the truck body. (Chuck Madderom)

Atlanta Fire and Rescue at Hartsfield-Jackson Atlanta International Airport displays its fleet of Oshkosh Striker 3000s. This is the mid-range Striker model, which has three axles. Striker models take their designations from their water tank capacity. The smallest is the Model 1500 two-axle unit, while the largest of the trio is the four-axle 4500 model. The 3000s seen here also carry 420 gallons of foam, 500 pounds of dry chemical, and 460 pounds of Halotron 1. (Oshkosh Corporation)

Oshkosh Corporation's all-black Velocity demonstrator pumper is billed as one of the safest fire trucks built. Its "one-eleven" mirrors meet visibility standards for school buses while a one-piece windshield with three wiper blades also increases visibility. Other safety features of the Velocity include a side-roll protection system, frontal crash air bags, and an automatic safety management braking system. (Oshkosh Corporation)

Rescue 5 of Milwaukee's main airport is a 2005 Colet Jaguar K-15S on a Marmon-Herrington chassis, made of stainless steel. The chassis incorporates a constantly self-leveling suspension system. The rig has a 50-foot articulated boom, 1,250-GPM Waterous PTO pump, 1,500 gallons of water, 100 gallons of AFFF, and 450 pounds of Purple K. Considered one of the most advanced fire vehicles, the Jaguar was first delivered to the U.S. Air Force in 2007. With a crew of two and a wheel-base of 180 inches, the Jaguar gets its name from its ability to accelerate to 50 mph in 14 seconds and to reach a top speed of 70 mph. A 500 HP Cummins ISM diesel engine supplies power. (Chuck Madderom)

A Pierce demonstrator pumper demonstrates its ergonomic "Pack Mule" hose bed, which lowers hydraulically to facilitate loading and unloading hose. Modern fire apparatus builders strive to increase firefighter safety and their workload. (Oshkosh Corporation)

New York's Ten Engine is a 2002 Seagrave rated at 1,000-GPM. The engine has multiple light bars, front and rear, and wears messages on its cab supporting troops and memorializing members lost on 9-11. (Author's Collection)